Rejuvenation in Fruit Trees

NIPA® GENX ELECTRONIC RESOURCES & SOLUTIONS P. LTD.
New Delhi-110 034

About the Author

Dr Rajesh Kumar, Principal Scientist at ICAR-National Research Center on Litchi, Muzaffarpur, Bihar, has vast (more than 27 years) experience in research and development and teaching in Horticultural Science. Dr Kumar started his career as Junior Scientist-cum-Assistant Professor (Hort.) at the Birsa Agricultural University, Ranchi, Jharkhand. He has tremendous research experience in various horticultural crops. He has developed many technologies with success stories. He has two tuber crop varieties to his credit. Since one and-a-half decade he is continuously researching on perennial fruit crops. Apart from this he has guided 4 M.Sc. students and remained member of advisory committee for 2 Ph.D. students in Horticulture.

He has published 45 research articles in Journals of National and International repute, presented papers in various (54) National and International seminars, symposia and conferences. He is life member of 21 Agri./Horti. societies and recipient of so many awards (recognition award, fellowship, best paper presentation awards and scientist of the year - 2014). He has written two books, 7 technical and extension bulletins, 19 book chapters and 66 popular articles including success stories.

Rejuvenation in Fruit Trees

(Tropical and Subtropical)

Rajesh Kumar
Principal Scientist (Horticulture)
ICAR-National Research Center on Litchi
Muzaffarpur - 842 002, Bihar (India)

NIPA® GENX ELECTRONIC RESOURCES & SOLUTIONS P. LTD.
New Delhi-110 034

NIPA® GENX ELECTRONIC
RESOURCES & SOLUTIONS P. LTD.

101,103, Vikas Surya Plaza, CU Block
L.S.C. Market, Pitam Pura, New Delhi-110 034
Ph : +91 11 27341616, 27341717, 27341718
E-mail: newindiapublishingagency@gmail.com
web: www.nipabooks.com

For customer assistance, please contact
Phone: + 91-11-27 34 17 17
Fax: + 91-11- 27 34 16 16
E-Mail: feedbacks@nipabooks.com

ISBN: 978-81-19254-22-4

Composed and Designed by NIPA®.

भाक्र अनूप
ICAR

भारतीय कृषि अनुसंधान परिषद्
कृषि अनुसंधान भवन–II
पूसा, नई दिल्ली–110 012

INDIAN COUNCIL OF AGRICULTURAL RESEARCH
KRISHI ANUSANDHAN BHAWAN-II
PUSA, NEW DELHI-110 012

डॉ एन.के. कृष्ण कुमार
उप महानिदेशक (बाग. वि.)

Dr. N. K. Krishan Kumar
DEPUTY DIRECTOR GENERAL (HORTICULTURAL SCIENCE)

Foreword

The horticulture sector in India has been changing very fast in the recent past, thanks to agroclimate, policy planning and scientific support, and hard-working farmers. The country has made remarkable progress in evolving varieties suitable for various purposes, identifying rootstock, integrated nutrient and water management, pests and disease management technologies. Emphasis on quality seed and planting material production has largely contributed in increasing productivity of horticultural crops. We all know that majority of the fruit crops are perennial and of high-value, in which meticulous planning for orchard establishment and subsequent scientific management of bearing orchards is required for getting optimum production.

Optimum canopy development through appropriate pruning and training is a pre- requisite for enhanced utilization of applied inputs and solar energy. During the course of continuous growth and production, the canopy of even the best managed perennial fruit trees becomes unproductive. Rejuvenation of such orchards through pruning and training, and adoption of improved input management practices like nutrient, water, pests and diseases management improve the vitality and can bring them back to economically-viable productivity level. Concerted research efforts have resulted in standardization of rejuvenation technologies of important fruit crops for various agroclimatic regions. There was a long felt need to compile the available information to present in a single volume like this for the larger benefit of stakeholders.

It gives me immense pleasure on going through the book **Rejuvenation in Fruit Trees (Topical and Subtropical)**, by Dr. Rajesh Kumar, Principal Scientist (Hort.), ICAR-National Research Centre on Litchi, Muzaffarpur, Bihar.

The book, covering 17 chapters including prospects and constraints of rejuvenation technology and practical application of rejuvenation packages in 11 high-value commercial fruit crops provides information on core areas of rejuvenation. I compliment the author for bringing out the publication and hope that the scientific information provided in the book will be greatly useful to stakeholders.

(N. K. Krishna Kumar)

Preface

India is the largest producer of many fruit crops in the world. Among them, perennial fruit crops have immense scope in domestic and export market. The country also provides suitable climatic conditions for these fruit crops, but the productivity is low due to interactive effect of many dependent and independent factors. The lack of awareness and required knowledge about the successful production system of fruit crops of perennial nature lead to push them in unproductive and declining phase. The old senile unproductive trees which become problematic to the orchard growers, need to uproot and replace them with new plantation, which is a long-term venture. The developed rejuvenation technology and its successful implementation, can transform the old senile unproductive phase to young productive phase in a sustainable manner through the integration of medium-term strategic plan in a scientific and skilful manner by utilizing the benefits of existing rhizosphere, without any economic loss. This vital aspect of rejuvenation in old unproductive fruit trees, presents many (a variety) cross-cutting issues on the theme of restoration of youthful vigour, enhancement of quantum of quality fruit production, period of sustainability and above all the economic feasibility. This necessitated that a comprehensive book on this aspect be written, which has collective information with critical reviews and practical concerns to avoid the hitch at initial level of adoption for the invasive rejuvenation technology for old senile fruit trees. With this idea in mind, the present book '*Rejuvenation* in *Fruit Trees* (*Tropical and Subtropical*)', would be highly useful to students of Agriculture/Horticulture, at B.Sc, M.Sc. and Ph.D. levels, academic experts, researchers, scientists, administrators, policy-makers, farmers, exporters, entrepreneurs and other stakeholders in fruit production.

The book consists of 17 chapters. The first 6 chapters deal with the concept, cause, need and strategic steps involved in technique of rejuvenation, while the rest 11 chapters takes care of high-value commercial fruit crops of woody perennial nature describing present status, various prospects and constraints, involved implementation strategies with integration of multipronged approaches where aftercare practices including nutrition, irrigation, canopy management have been given proper attention along with damages caused by disorders, pests and diseases. The current trends and requirements have also been brought within purview in addition to critical perspectives that foster a better understanding of its R&D issues for this technique, which is based on possible updated information vis-à-vis the long research and field experience of the author.

The author acknowledge the help of scientists, experts and research workers in consulting their opinion, reports, research papers, books, bulletins etc., published in India and abroad. The author is also thankful to those fruit growers, where the successful implementation of the technology and success stories gave the courage to write this book. I am more thankful to Dr. Som Dutt, Editor, *Indian Journal of Agricultural Sciences*, and *Indian Horticulture,* DKMA, KAB-I, New Delhi, for critically editing and preparing the manuscript in an easily readable mode. I express my sincere gratitude to my parents and family members, who continuously encouraged in this venture of writing the book. There remains all the possibility of further improvement and betterment, so I also urge upon the readers to share their critical views/reviews/opinions on my email: rajeshkr_5@yahoo.com : for correction and modifications in the future editions.

Muzaffarpur, Bihar **Author**
Date 31-12-2015

Contents

Chapter 1

Introduction

India is endowed with abundant natural resources and diverse agroclimatic conditions, which are ideal for cultivation of a number of annual or perennial crops, in different areas. Meticulous planning, extraordinary scientific support, judicious use of inputs and focused policy thrust have led to manifold increase in production and productivity of fruit trees. The horticultural crops, particularly fruit crops, are high-value and technology- intensive crops. Appropriate technologies are gaining popularity in entire country in view of their high potential of productivity, land utilization and more income. Today, our country is producing different kinds of fruits, but quantum of total fruit production is unable to cope with the present day demand. Various constraints, biotic and abiotic in nature, pose hindrances to meet the fuel potential. The poor productivity of fruit trees is the main reason of low availability of fruits. The low productivity of fruits is compounded by several environmental, genetic and conditional factors. Besides, productivity and quality of perennial fruits continued to remain low and there exists a wide gap in between actual and potential production. Visualizing the problems of perennial commercial fruit production, the main reason comes in forefront/top is a large number of old and senile orchards turning even unproductive, due to lack of information regarding non-adoption of appropriate agro-techniques and canopy management (Hare Krishna, 2012). Though, we have made rapid strides in fruit production, the productivity is very low compared to developed countries. The present scenario of production in our country is still technology deficit and what so ever the technologies are available very poor in adoption level and the situation still go worse even in the era of climate change.

Present situation demands that new technologies are needed to push the yield frontiers further by utilizing inputs more efficiently, diversifying to more sustainable and yielding higher quantum of quality production, particularly by integration of invasive and non-invasive medium-term strategic technologies. There are still challenges ahead in the way of the research which require concerted efforts in organizing, conducting and developing the technology as well as the proper way they are transferred to growers. The new generation of technologies will have to be much more site-specific, based on high quality science and a heightened opportunity for end-users, participation/adoption in identification of targets and achieving the objectives. These must be not only aimed at increasing the growers/farmers technical knowledge and understanding of science-based timely operations. It will also need to take on challenges of integrating the socio-economic context and role of markets for profitability and economic viability.

It is a reality to see that every living organism after some period of time (that may vary from species to species) loses its efficiency to perform various live and physical functions. Likewise in fruit plants also, there is a decline both in quality and quantity of produce after some period of time. As a result of which orcharding becomes economically non-viable and non-remunerative. In the recent past, declining productivity of old and dense orchards existing in abundance has become a matter of serious concern for the orchardists, traders as well as scientists.

The progress of science and technology no doubt is based on imagination, desire and dream and the resultant progress made mainly helped the mankind to improve the quality of life along with its allied product and produce. Though processes and procedures require scientific skill and refinement, but look, where we will, we find science and technology helping us in countless ways. So, if we have determination, we can think of transforming the phase according to will.

In India, a considerable area (about 30-35%) under fruit crops is occupied by old, dense and diseased orchards. For overcoming the problem of unproductive and uneconomic orchards existing in abundance, large-scale uprooting and replacement with new plantations (rehabilitation) will be a long-term and expensive strategy. Therefore, research efforts are required to be initiated to develop and standardize a technology for restoring the production potential of existing plantations by a technique called rejuvenation.

Chapter 2

Present Status and Production Constraints

Today, our country is the second largest producer of fruits in the world, yet the per caput availability of fruits is below the recommended level (Chadha *et al.*, 2013). The major reason for low availability of fruits is poor productivity. In addition, pressure of burgeoning population and ever increasing demand also result in reduced availability of fruits. The production of fruit crops is still nature dependent. The problem of low productivity of fruits is compounded by several environmental, genetic and conditional factors. At present, there exist a wide gap in between actual fruit yield and their potential yield in almost all tropical and subtropical fruit species, particularly in perennial and woody trees of fruits.

There is a great diversity in perennial fruit crops which permits considerable manipulation for various purposes and desirable outcome. For achieving even the average productivity in many fruit crops, the production constraints like lack of information regarding canopy management, non-adoption of appropriate agro-techniques and timely management practices are required, while the concern about existing a large number of old and senile orchards which has turned or are turning even unproductive needs attention (Hare Krishna, 2012). Though, the country's fruit production status is showing an increase, but far below the potential production requires immediate attention to critically analyze the situation by revisiting the problems and its causes, identifying the gaps and development of need- based technologies in the present context.

The present status indicates that fruit production over the past few years, has been showing an increasing trend in our country but the low productivity and that too of poor quality dominance in some important fruit crops of high commercial value, more particularly of perennial in nature is causing concern. Among major perennial nature of fruit crops, mango, litchi, guava, citrus, aonla (Indian gooseberry), cashew and pomegranate are important among tropical and subtropical groups. These fruit crops are very much associated with livelihood security of people in the region whereever. They are grown in abundance.

Although, India is a leading mango producer globally, with an annual production of 15.19 million tonnes from 2.29 million ha, represents 36.0 % of the total area under fruit production, with average productivity of 6.8 tonnes/ha, which is quite low as compared to other mango-producing countries (Mishra, 2013). In litchi, our country is the second largest producer after China in the world. It is cultivated in 82,700 ha with a total production of 5,80,100 tonnes, and productivity of 7.10 tonnes /ha (NRC on Litchi 2014). In India, area under guava is estimated to be more than 2.05 lakh ha, producing 2.46 million tonnes with a productivity of 12.00 tonnes/ha and this is fourth most important fruit crop in the country (Mitra,2008; Singh *et al.,* 2013; ICAR Databook 2014).

Aonla is an indigenous fruit of Indian subcontinent and ranks first in area and production. It is growing in 49,620 ha, with a total production of 1,50,000 tonnes (Pathak, 2003; Singh and Mishra, 2007). The yields often reflect irregular annual bearing patterns even with poor quality. They vary greatly from season to season, apart from the other associated problems. Cashew is being grown in 9.91 lakh ha with an annual production of 6.92 lakh tonnes in our country. The national productivity comes to 7.49 q/ha (Hubballi and Jnandevan, 2012) against the target of 20.00 q/ha (Dora and Sethi, 2013). There is observed decline both in quality and quantity of produce after some period of time, making orcharding economically non-viable and non-remunerative. In India, citrus fruits are grown 9.88 lakh ha with an annual production of 100.90 lakh tonnes (ICAR Databook 2014). Different types of citrus species have different region-specific dominance of quality production. Pomegranate is also one of the important, high priced tropical/semi-arid fruit crop grown in India and its annual fruit production is 7,45,000 tonnes from 1,13,000 ha, the average productivity of 6.6 tonnes/ha only (ICAR Databook 2014), which is quite low as compared to other countries (Hiwale, 2009).

Chapter 3

Cause of Un-productiveness and Decline

The low productivity of perennial fruits has been attributed to various factors, like small size of holdings, preponderance of old and senile trees and poor management of inputs such as water, nutrients and pesticides. Thickly-shaded mango orchards in Malihabad areas of Uttar Pradesh (Kalloo *et al.*, 2005; Mishra, 2013), lanky as well as sick tall unproductive litchi orchards in Muzaffarpur and its adjoining areas in Bihar (Kumar, 2008), seedling orchards of guava and citrus fruits throughout the country, including disease-affected aonla and other perennial fruit plantations are commonly seen in large tracts in different parts of the country (Pandey *et al.*, 2013; Singh 2007). In pomegranate, the old orchards become non- productive due to incidence of pests and diseases, and negligence (Hiwale, 2009).

The traditional trend of orcharding for most of the tropical and subtropical evergreen perennial fruit trees, in general, are regarded as not responsive to pruning. However, a number of observations made in recent decades have conclusively proved the effectiveness of this important operation in maintaining the sustained fruit yield and keeping the plant well within manageable limits. The efforts have been made to deal perennial fruits in quite details wherever deemed necessary. Climatic factors have been found to affect significantly on fruit bearing as a result of pruning treatments. Though, this aspect is one of the most misconceived and least understood when seen with shoot emergence and shoot physiology for perenation and phase change from vegetative to reproductive. Correct pruning may easily spell the difference between success and failure, yet many growers undertake infinite pains to get the best out of their trees but the neglect to give their consideration to important practice. It is obviously needless to mention here that their such efforts end in futility. For better understanding of principles and practices of pruning, it would be imperative to trace out its historical background. The present level of understanding about pruning has been developed through ages with the increased understanding of growth and developmental processes of fruit plants.

In recent past, there has been an increasing interest to understand pruning practices, their principles and techniques. The improvements have been influenced by the ever changing economic status in fruit industry in the world. During the past few decades, large data have been added in the field of pruning, which have been responsible for a different interpretation of pruning. Later, fruit growers started giving prime consideration to their economic conditions, interest in adoption of some vital technologies increased, they started pruning/ training the fruit plants with respect to their fruiting capacities.

The great increase in production costs which occurred stimulated the growers to realize the need for greater efficiency of management to increase the yield of high quality fruit per unit area economically. The renewal system was thus rapidly adopted. Most of the modern training methods which have been evolved and protocols for that have been standardized. Likewise, modification continued as evolutionary process bringing about the improvements leading to the standardizations of this most important aspect of fruit culture.

The pruning has been dealt with details especially in respect of its requirement for individual fruit trees. The objectives, principles and the economics of pruning of various fruit trees have been comprehensively explained with real experimentations, practical examples, success stories and hints to meet the requirements of a variety of fruit growers engaged in fruit production.

In the recent past, declining productivity of old and dense orchards existing in abundance has become a matter of serious concern for orchardists, traders as well as scientists. According to Singh (2007), there are many causes which make the orchards uneconomical : i, growth of wild shrubs and grasses (as weeds); ii, unsystematic planting at initial stages; iii, inferior varieties; iv, heavy infestation of pests, diseases and parasites; v, damage due to adverse weather conditions, rodents and other enemies; and vi, old trees with advanced age have tendency to develop hollows in their trunks and main branches starting from crotch or from the point of bifurcation.

On the basis of extensive survey reports undertaken for many fruit crops revealed the fact that there is considerable increase in area under old senile orchards (30-35%) (Kumar, 2008; Lal *et al.*, 2000). It has also been observed that loss of tree vigour in older trees results in falling of foliage and low yields of small fruit, and is usually accompanied by dieback of twigs and small branches, risk in harvesting of fruits from upper canopy of lanky and height. This decline may be due to age, disease, soil pests, water induced root damage, and other causes. Old trees often produce fruits and are difficult to harvest due to their size. Moreover, old trees harbour pest and diseases and it is difficult to reach the whole tree, while spraying to control them. This chunk of old senile orchards, apart from being highly uneconomical, act as source of disease and pest infestation and their uprooting must be vigorously enforced by the state through new legislations.

Chapter 4

Need of Rejuvenation

Rejuvenation of fruit trees is visionary approach to save the old senile declining trees for future sustainable quality fruit production through unique integration of invasive exhaustive operations in scientific manner and skilful management. Since unproductive and uneconomic orchards are in abundance, a large-scale uprooting and replacement with new plantations (rehabilitation) will be a long-term and expensive strategy. Therefore, research efforts have been initiated at various pioneer institutes/ SAUs to standardize. The technology for restoring the production potential of existing plantations by a technique called rejuvenation. Rejuvenation of old orchard is very vital under orchard management. Reiterative pruning and canopy re-building of perennial fruit trees is medium- term strategy requiring scientific skill and technique, for bringing back the youthful vigour with enhanced quantum of quality production in sustainable way within a time gap of three years for mango, litchi and cashew (Lal *et al.*, 2000; Kumar, 2008), while two years for guava, aonla and pomegranate (Singh, 2007; Hiwale 2009) without any economic loss to growers. The old fruit orchards need to be rejuvenated since they show decline in yield and quality of produce. These trees/plants need to be rejuvenated which have attained a stage where they are no more profitable from the grower's point of view (Singh, 2007; Srivastava, 2007).

There are 25-30% fruit crop orchards which are old, senile and unproductive, while the quantum increase in present level of production through horizontal expansion (i.e. new plantation) will be a long-term venture. The existing orchards turned or turning to unproductive and uneconomical ones, can be brought back productive ones through rejuvenation, which in turn help in restricting the deterioration of health of plants and thereby extending their life, aims at improving the yielding capacity of quality production with renewable ability and situation under minimum period of time as compared to establishment of new orchard, which attains the commercial bearing stage after 8-10 years (Kalloo *et al.*, 2005; Lal *et al.*, 2000; Kumar, 2008; Singh, 2007; Hiwale, 2009; Pandey *et al.*, 2013).

Chapter 5

Physiology and Mechanism of Rejuvenation

Perennial fruit trees build their woody architecture with some building blocks, i.e. modular units, known as metamer or phytomer (White 1979; Room *et al.*, 1992; Hare Krishna, 2012). It consists of meristems initially an apex. When a meristem faces periods of dormancy during low temperature or pruning, later it appears as bud, whenever active, as such bud produces a growing point with sequence of meristematic tissues simultaneously called a sprout, i.e. extension unit or growth unit (Halle *et al.*, 1978; Bell, 1991). The driving mechanism behind the pattern appears to be auxin production (and basipetal transport) when new shoots and leaves developed and are exposed to light (Sterck, 2005). Other feature of trees are to produce more xylems relative to phloem, accumulate xylem over time and may thus produce thick woody branches and stems. Further, attaining reproductive phase and reproduction processes compete with vegetative functions, depleting resources necessary for maintenance and growth (Fakhri *et al.*, 1987).

The idea that reproduction, growth and defence interact within the individual and compete for limited resource is now considered an established principle. Because there are trade-offs between a plant's various functions and the concept of source, and sink helps to explain allocation patterns at both the physiological and evolutionary levels. Logical approaches have extended this hypothesis to an explicit as source and sink framework in which the effects of defence allocation on rates of consumption can be weighed against the effects on growth (Coley *et al.*, 1985; Mooney and Gulmon, 1982; Fakhri *et al.*, 1987). The allocation to vegetative structures has compounding interest for plant growth, as plant life-cycles begins with pure vegetative growth (Abrahamson, 1979), also consistent with the idea that perennials must forego some reproductive expenditure to retain sufficient resources for perennation.

Perennials have varying levels but in higher reproductive allocation (Hickman, 1977; Hare Krishna, 2012). The reproductive characteristics are among the most conservative traits in plants, but the prediction trends for reproductive allocation are less where morphological constraints on resource allocation are similar in case of perenation. Photosynthesis either directly or indirectly provides the building blocks for all biomass production and growth. However, plant growth and productivity are not determined alone by photosynthetic activity but also by the way, in which the product of photosynthesis are allocated and utilized during plant growth and development.

The decline in productivity of orchards is largely be due to poor photosynthetic efficiency besides several other physical compounding factors including age of the orchard, dense and intermingling branches, neglect and poor management, bearing poor quality fruits, and attaining unproductive stage. This requires formative or corrective pruning to reinforce the tree to solve the bearing problem of old orchards, i.e. making them again productive. It needs adoption of rejuvenation technology (Kalloo *et al.*, 2005; Kumar, 2008; Singh, 2007). The technology requires reiterative pruning, which is of invasive nature, because fruit trees are often lopped extensively through skeletization to encourage further re-growth, simultaneously, causing a growing sense of anxiety to farmers that their trees eventually die. However, hardly reports are available about the death percentage of trees subjected to such a severe salvaging / resuscitating treatment going for rejuvenation, if followed scientifically. Some studies suggest that tree survivability to frequent pruning may be related to tree age and relationship between age of tree and pruning intensity on survival and biomass production (Duguma *et al.*, 1988; Keerthisena, 1995; Tipu *et al.*, 2006). Although the elucidation of this relationship remains unrealised, some insights in research results have been provided but still needs scientific studies and pursuance.

After severe pruning, the leftover thick branches of trees are the primary source of reserve carbohydrates in the absence of shoots and leaves (Hare Krishna, 2012). Here, rejuvenation refers to the process of change, which occurs during the reversion from old to adult to juvenile (Hare Krishna, 2012). The appearance of sprout, giving rise to shoots and young leaves emerging on pruned/cut branches of old trees depend on imported carbohydrate (sucrose) from sieve tubes of phloem and other tissues to provide metabolism and carbon skeletons for metabolic inter-conversions and biosynthetic process. The pathways that mediate carbohydrates regulation in pruned trees have been studied in some perennial crops. Vegetative re-growth as sprout emergence is a sink for re-sources, and so allocation to it must be under ecological controls similar to those that govern sexual reproductive allocation. The effects of specific types of stress seem to vary widely, but clonal growth does appear to be tied to resource availability (Ashmun and Pitelka, 1984) and plant size (Grace and Wetzel, 1981).

Despite the widespread occurrence of this physiological integration in emergence of vegetative sprouts, our understanding of ecological and evolutionary, source and sink study is still minimal. Debate continues over whether re-growth represents reproduction or only vegetative growth and perenation continues taking a certain gap period for phase conversion (Kumar, 2012). Phytohormones regulate maturation in plants and ratio of abscissic acid, gibberellins, auxins and cytokinins, rather than their absolute concentrations play a primary role in causing maturation, as these phytohormones probably do affect changes in maturation state and the pruned plant successfully integrate several environmental signals to undergo developmental transition (Mallik and Srivastava, 1985; Singh, 2003; Hare Krishna, 2012). Until vascular connections to new sprouts, vegetative re-growth are provisioned during adverse conditions.

There is existence of meristematic tissues in the leftover trunk portion after heavy pruning, which are capable of active cell division and differentiation into specialized tissues as shoots and leaves at initial stage importing stored carbohydrates (Singh, 2003; Hare Krishna, 2012). Emergence of vegetative sprouts, leading to shoots, becoming older as stem and branch but does not possess flower buds (panicles) for one or two years. The developing new shoots (pre-formed growth) of current season, continues growth, taking enough time to mature to go for phase change. It experience dormancy period due to winter exposure (low temperature) for one or two seasons are probably changing to floriferous condition, as its vegetative meristem has differentiated into inflorescence (Singh, 2003; Hare Krishna, 2012; Kumar 2012). Since, flowering in perennial fruit crops is complex and cannot be described as a single trait, physiology and molecular mechanisms involved in the integration and regulation of some perennial specific traits, which prevent floral induction until certain developmental stage has been reached, thus ensuring that the plant has sufficient resources to be able to sustain flower and subsequent fruit production (Hare Krishna, 2012).

It is important to develop a framework sufficiently strong to bear the weight of quality fruits The consequences of severe pruning, method as well as time gap of rejuvenating old trees along with cultural practices required to support the reinvigoration process (Kumar, 2012). Thinning of shoots after six months of growth, along with selection and allowing of the primary scaffolds to be properly trained outward to grow fast. Scaffolds should be headed during the dormant season of first two years to promote continued lateral branching to be stiffen and strengthen, as well as for maturing the scaffolds for shaping desired framework (canopy). Often interactions between environmental stimuli and endogenous developmental factors exert some regulatory effect, causing phase change to floriferous condition (Hare Krishna, 2012). Floral induction in some perennial fruit crops (mango, litchi, guava, some citrus etc.) occur in response to cool temperatures perceived by mature leaves largely regulated on physiologically mature shoots (Menzel and Simpson, 1995; Menzel *et al.*, 2000; Hare Krishna, 2012; Kumar 2012). Contribution of only some percentage of primary flushes in blooming rather than whole still needs further investigation which may involve biochemical and hormonal assays of shoots (Hare Krishna, 2012; Kumar 2012).

There is a need of periodic monitoring of pruning operation, emergence of re-growth and its maintenance avoid complexities, and disappointments due to undesired framework development or chances of mortality (Kumar 2008). Although genetic re-combination does not occur in production of new sprouts in meristems and vegetative growth does serve many of the same functions of reproductive phase after attaining perennation and maturation and further in sustainable manner. The physiology and mechanism behind to make the older trees perform parallel to those of well-maintained adult trees or even more still require experimentation at precision level to deal with more authenticity in the form of recommendation (Kumar, 2009, 2013).

Chapter 6

Technique of Rejuvenation

There is an urgent need to take up productivity improvement programmes in the senile plantations, with fresh stock supported with appropriate and integrated combination of inputs, pruning and grafting as well as training techniques taking care of health, vigour and productiveness. This require proven technology implementation through individual farmers, farmers' cooperatives, self-help groups, NGOs, growers' associations and commodity organizations. However, technology depends upon the nature and requirement of a particular fruit crop. The rejuvenation technology for transformation of old senile orchards into young productive ones in sustained manner requires well-planned and concerted efforts. Though processes and procedures require scientific skill with precision in application/implementation as these important commercial crops have been associated with livelihood of many farmers/growers, i.e. why economic production is of utmost important for the growers as it holds true for all the region specific important commercial fruit crops in general.

The successful translation of rejuvenation technology in old senile orchards includes sequential operational steps. They should be followed in scientific manner for different category/group of fruit crops.

Identification/Selection of Old Orchards

The cause of decline should be determined and corrected if possible; otherwise, response to rejuvenation pruning is temporary. If trunk and basic scaffold limbs of trees are not structurally sound due to disease, heart rot, and cold cankers, such rejuvenation procedures are not justified. A limited root system due to a high water table, a hard pan or poor subsoil can result in early tree decline with trees being more readily stressed by drought. When this occurs, the top becomes out of balance with more limited root system with consequent insufficient uptake of water and nutrients by roots. Topping these trees should temporarily restore them to a more favourable top to root ratio, but not alleviate the long-term problem.

The severity of pruning for rejuvenation depends on the cause and degree of decline. It has been observed in general that orchard's dense, compact top canopy cover with bare (devoid of leaves), long-high braches at the bottom and bearing fruits only on high-tops posing a lot of socio-economic problems to growers. It becomes uneconomical to maintain after commercial bearing age/ stage as specified for many commercially important fruits crops. The yield loss is also caused due to bearing of poor quality fruits at high up canopy, difficulty in harvesting and also due to physiological disorders, incidence of pests and diseases, lack of resources and facilities and developing tendencies of alternate/ biennial bearing. It also poses problems in proper management. Such orchards need to be rejuvenated for further higher production of quality fruits. But it is important to look for the health, location and value of fruit trees, which should be determined before going for rejuvenation (Kalloo *et al.*, 2005; Lal *et al.*, 2000; Kumar, 2008; Singh and Mishra, 2007). Sometimes, condition of orchard is so poor, some corrective measures are required to be applied before going for rejuvenation technology. Then only, it should be emphasized finally to go for the programme of reiterative pruning and efforts should be made for satisfactory growth maintenance by other required cultural and management practices (Singh, 2007).

Training to Orchard Owners/Awareness to Growers

The initial level of adoption, invasive rejuvenation technology for old senile fruit trees, presents many (a variety) cross-cutting issues on the theme of restoration of youthful vigour, enhancement of quantum of quality fruit production, period of sustainability and above all the economic feasibility (Kumar, 2008). It encompasses aspects of scientific suitability, productivity, sustainability and economic feasibility. All these require proper implementation leading to success stories. This concern for the awareness to growers and enough knowledge base through training, convincing the economic feasibility is very vital to increase the level of adoption (Kumar, 2008).

For better understanding of principles and practices of intensive hard core pruning, it would be imperative to trace out its historical background of growth and developmental processes of fruit plants for the required pruning practices, their principles and techniques to be adopted. During the past few decades, large data have been added for the operation of light pruning, reiterative pruning and training to enhance the production efficiency of orchards which are to be integrated for purposeful interpretations.

Considering various pros and cons of rejuvenation, it will only be programmatic to undertake this programme after giving proper training for acquiring knowledge to orchardist to develop the expert system and apply the methodology in a scientific manner. It could be elaborated that the fruiting potential of trees is largely governed by its architecture, canopy density and photosynthetic efficiency. Decline in productivity of orchard could largely be due to poor photosynthetic efficiency besides several other compounding factors including age of orchard, dense and intermingling branches, neglected and poor management of orchard. Productivity status is further affected due to various problems of insect, pests and diseases, which perpetuate in abundance in such orchards, should be properly dealt.

The regular programmes to determine the growers, specific needs, attitudes and traditional knowledge of crop, their views, which can be later taken into account in the implementation of planned activities. The use of different communication approaches such as mass media, meetings, seminars and training sessions at different levels, personal contacts, campaigns, and field supervisory visits greatly improved coordination, linkages and team work spirit in implementing this type of medium-term planned strategic activities and operations, which are mainly invasive in nature.

Reiterative Pruning

There is a need of knowledge concerning the rejuvenating old senile trees with proper techniques. Pruning plays an important role in regulating and controlling growth, flowering and fruiting in perennial fruit trees (Crane *et al.,* 2009; Singh *et al.*, 2012). In high-density planting, trees are pruned regularly to control size and shape (Iyer and Kurian, 2006). Different systems/patterns of hedging or pruning are used. Rejuvenation is vital to a tree's health, especially when our goal is to produce enhanced quantum of high quality fruits over a number of years in a sustained manner by transforming the old senile unproductive trees with the help of proper pruning and training techniques. The principle is creating a crown on dwarf fruit tree by transforming size of old and lanky (tall) fruit trees.

Many mistakes are made when people prune large or heavy branches. In many cases, pruning process often results in damage to the trees. Pruning of large or heavy limb (any limb greater than 2 inches in diameter) should be steps process in order to avoid tearing or ripping of bark while making the cut. When single cut in a hurry is made, the weight of branch or limb may cause the limb and bark to tear several feet down the trunk before the cut is completed. The step-wise reiterative pruning cuts should be made as the first cut is made underside or bottom of the limbs/branches, approximately one-fourth to almost half way deep. The second attempt of pruning cut is on the upper side top of the limb, just opposite or may be slightly outer side but in no case underside or bottom side of the trunk. The final and finishing cut is made at the trunk marked at the desired height, since the weight of limbs/branches has been removed, the final cut can be made with precision and without the risk of damage to the bark of the trunk. Use of pruning paints and dressings at the cut/wound portion is very important to act as defences against disease and pest infestation.

When deciding which branch to cut and where to cut it, remember that topping a vertical branch encourages vegetative growth necessary for development of tree and opens the tree to more sunlight. Topping horizontal branches is done to renew fruiting wood and to thin off excessive fruits. Horizontal branches left uncut bear earlier and heavier fruits. Upright branches generally remain vegetative and vigorous. Horizontal branches generally are more fruitful. A good combination of both is necessary, for fruiting now and in future years. Remove suckers, water sprouts and most competing branches growing straight up into the tree. Downward bending branches eventually lose vigour and not contribute to the canopy, produce almost no fruits; cut off the part hanging down. Do most of the pruning in the top of the tree so that lower branches are exposed to sunlight. Sun exposed wood remains fruitful and produces the largest fruits. Shaded branches eventually stop fruiting and never produce without drastic topping and renewal of the entire tree.

Make clean cuts without leaving stubs. In abandoned or neglected fruit trees, which have started declining or turned unproductive, pruning may look like a difficult task. In such cases, these trees can be reiteratively pruned in a phased manner and made functional carefully in the orchard. Although trunk and general framework may be sound, the functional portion of tree is usually a solid canopy of weak, crowded branches at the top or periphery of tree's canopy. Trees may have annual shoot growth, properly trained with framework. Although pruning is one of the oldest horticultural practices, the principles on which it is based are not always understood. The novice tends to focus on minor details at the expense of some well-accepted principles. Whether the task is to prune neglected, mature trees or simply annual pruning of trees under regular care and maintenance, the certain laid out principles should be taken care off.

In rejuvenation, heavy reiterative pruning of almost all perennial fruit trees is done at certain effective height of 1.5/3 m depending upon the nature and growth of fruit crops as well as girth and type of main trunk. About 2-3 main branches with outward growth from the base are marked for pruning at required/ recommended height, with the plan of developing umbrella like or semi-circular framework of tree canopy to enhance bearing surface area. Reiterative pruning can be done either with manual saw or power-operated saw in a phased manner from the top.

Time of Reiterative Pruning

In principle, prune perennial fruit trees when they are in dormant phase. It is usually done just after harvesting. By adopting and following all the required scientific procedures, the best time to go for reiterative pruning for rejuvenation is also the vital aspect. Rejuvenation is labour intensive operation and it is extremely necessary to complete the operations in time for achieving the desired results. It is also to be aware about the time when not to go for rejuvenation pruning. The misbalanced root:shoot ratio can be corrected by judicious pruning. The branches which have died, broken or one which interferes with natural growth, and water sprouts should be removed any time but not before flowering period. If the exposed wound are big enough, they should be disinfected with arsenical paste and treatments time to time. The time of reiterative hard core pruning in case of perennial fruit crops mainly done at the end of active vegetative phase or main dormant phase of annual production cycle depending on the crop and cultivars (Table 1).

Table 1. Recommended height and time for rejuvenation of old senile orchards of perennial fruit crops

Fruit crop	Recommended height for pruning (m)	Time of pruning	References
Mango	3.0-5.0	November-December	Kalloo *et al.* 2005
Litchi	2.0-2.5	August-September	Kumar 2008
Guava	1.0-1.5	May-June	Kalloo *et al.* 2005
Citrus	1.0-1.5	May-June	Kalloo *et al.* 2005
Aonla	2.0-2.5	December-January	Kalloo *et al.* 2005
Cashew	1.0-1.5	May-June	Kalloo *et al.* 2005
Pomegranate	1.0	May-June	Hiwale 2009
Sapota	1.0-1.5	August-September	
Ber	1.0-1.5	May-June	
Bael	1.0-1.5	November-December	

Source : Kumar 2015

Rejuvenation by Top-working

Top-working can be easily adopted in pruned trees to upgrade seedling plantations and inferior varieties of some fruit crops (mango, litchi and cashewnut) with superior commercial cultivars or newly-introduced selections/hybrids suitable for domestic and export markets. Scions of desired variety can be grafted on the newly emerged shoots developed on branches of pruned trees by adopting veneer or soft wood grafting during July-August with all care and procedural precision. Only 4-5 grafting per branch/trunk should be performed. For budding operation, scion shoot should be selected from the mother plants, which are desirable and prolific-bearer and free from disease and pest incidence. The scion shoots of superior variety can be budded on newly-emerged shoots developed on beheaded branches by adopting T-budding in citrus and patches/ modified ring budding during May-September in aonla. After successfully top-worked trees side shoots which emerge on pruned branch should be removed regularly as and when they emerged, so that the tree of superior variety is obtained.

The old unproductive and inferior seedling trees (mango, aonla and cashewnut) which are found in large numbers can be rejuvenated and improved by the process of top-working. The selected scaffolds limbs of inferior or unproductive trees, desired to be top-worked are headed back during February-March. The cut ends are treated with Bordeaux paste. Many shoots emerge within a short time below the stubs. Out of these, 2-3 vigorous shoots per branch or limb are selected and remaining are removed. These shoots make fast growth and become suitable for grafting by August-September of the same year. Select the best sprouts on each stub and graft them with the desired cultivar. Remove the rest of the sprouts gently by giving a clean cut close to the stub so that these may not re-sprout. If any of these shoots are left they will overpower the newly sprouted grafts. This way one has 8-10 grafted shoots. These grafts sprout within 20 days of grafting. The shoot above the sprouted graft should be headed back by keeping only one whorl of old leaves in November. These leaves provide food to the newly sprouted graft. Newly sprouted grafts should be covered with rice trash or plastic bags to save the grafts from frost or severe winters in December. The covers may be removed during February-March. While removing the covers, polythene sheet used for tying the graft union should also be gently removed. The old sprout above the graft union should also be re-cut, keeping only the newly sprouted on grafts.

Care should be taken that grafts should not get damaged since these cannot tolerate even small pressure and even jerk. For at least two years take care that no sprout should come up on the tree limbs other than the grafted shoots. Top-worked trees come into bearing within three years, depending upon the grafted cultivar (Singh and Mishra, 2007). Top-working is also used to rejuvenate unproductive and mature cashew trees from 5-20 years of age. The reiteratively pruned tree's trunk/branches, the emerged shoots should be grafted with scions of high-yielding cultivars using the soft wood grafting technique. The top-worked trees start giving fruit from the second year and third year from initiation of reiterative pruning operation (Duke,1989; Janik and Paull, 2008). The side grafting technique of two scions combined with two productive branches remaining on tree proved to be the quickest and most effective technique to rejuvenate cashew trees (Suharto *et al.*, 2012).

Application of Manures and Fertilizers

Non-application of nutrients at appropriate time with inadequate quantities was also one of the important reasons for reduction in productivity and development of hunger or deficiency symptoms in orchard making unproductive and poorly developed tree stature. There should be integrated nutrient management for maintenance of soil fertility and plant nutrient supply to an optimum level for sustaining the desired growth and further productivity level. As such, it is necessary that fertilizer application is made using both organic and inorganic in pruned orchard with the main intention to boost initiation of new shoots with vigour at an early date. After the reiterative pruning, the tree should be fed with optimum fertilizer doses, followed by irrigation near the root zone, just like the commercial bearing orchards.

The manures and fertilizers should be applied through ring/trench method. Soil properties, nature of crop and management practices greatly influence the availability of soil micronutrients to plants. To know the extent of micronutrients deficiencies in soils, application of straight micronutrients (single element) and fertilizers is more beneficial. Further, soil application of zinc and boron is found much efficient in correcting deficiencies and leaves residual effects for succeeding production, while foliar sprays also proved efficient and economic too in emergent conditions. Therefore even in cases of acute single element deficiency, use of multi-micronutrient mixture may be beneficial in yield and quality of fruit production.

Thinning and Canopy Development

After pruning to improve productivity of old and unproductive perennial fruit orchards, rebuilding canopy is very vital (Srivastava, 2007). The reiterative pruning for top off, brings invigoration effect and give rise emergence of numerous shoots on the entire body of trunk requires proper thinning. Thinning of shoots and canopy development should be skilfully done by just removing the extra, overlapping, down growing, week shoots after a certain gap period (of at least six months for mango, litchi, aonla and cashew, while one month for guava, citrus and pomegranate), then operation is repeated 3-4 times with the same time interval but just before the onset of rainy and winter seasons (Table 2). The enormous number of branches come out and cover the entire body of the pruned plant. Now, it is required to thin out the excessive branches as well and keeping only 3-4 number nearly top-side portion on each pruned limbs of the trunk projecting all the directions and further training for acquiring desired shape.

The thinning operation is required to be done carefully with the ultimate aim of developing solid, semi-circular or open umbrella type canopy. After this operation, there is generally a rapid period of shoot elongation and leaf expansion, followed by a period of leaf maturation before the next period of shoot growth. The duration and interval of growth are strongly related to temperature and irradiation (Batten and Lavah, 1994; Olesan *et al.*, 2002). The air and light must have access to all the parts of the tree. In guava and citrus, overlapping of fruit growth developed as a result of pruning, makes it possible to harvest 3 crops in a year (particularly in guava), with proper management of new shoots and flower initiation. The young shoots can be forced again to flower and fruit in the next season. The success of this technique depends upon the proper management of shoots through precise and timely pruning.

Table 2. Recommended time gap for starting of thinning and canopy development operations in rejuvenated old senile orchard

Fruit crop	Time gap for thinning (months)	Remark	References
Mango	6	-	Kalloo *et al.*, 2005
Litchi	6	-	Kumar, 2008
Guava	1	Concurrent for new fruiting shoots	Kalloo *et al.*, 2005
Citrus	1	Concurrent for new fruiting shoots	Kalloo *et al.*, 2005
Pomegranate	1	Concurrent for new fruiting shoots	Hiwale, 2009
Aonla	3	-	Kalloo *et al.*, 2005, Singh and Mishra, 2007.
Cashew	3	-	Kalloo *et al.*, 2005
Sapota	3		
Ber	3		
Bael	3		

Source : Kumar 2015

The tree training is required for canopy re-building to make a semi-circular / open umbrella shape canopy with enhanced surface bearing area (Kumar, 2008). Strategic thinning and pruning help to control tree size but it must be carried out at the right time to ensure shoot maturity, panicle emergence, flowering and cropping (Olesan *et al.*, 2002; Kalloo *et al.*, 2005; Kumar, 2013). Pruning and training of shoots of canopy outer surface is very important for shoot maturity and phase change from vegetative to reproductive stage in the third year after the reiterative pruning operation. Many subtropical and tropical evergreen fruit trees (litchi and mango) grow by recurrent flushes (Olesen *et al.*, 2002) can flower profusely but set relatively a few fruits (McConchie and Batten, 1991; Anila and Radha, 2003) and large trees with a high leaf area to canopy surface area ratio (Menzel *et al*., 2000).

There have been relatively few studies on the response of tropical and subtropical evergreen fruit trees to pruning, in contrast to the abundance of studies for deciduous species (Mika, 1986; Singh, 2010; Srivastava, 2007). Since cashew produces fruits only in the periphery of the canopy, massive supporting branches beneath is of little value as far as increasing productivity is concerned, hence pruning technology coupled with thinning and training attempts are made to develop enhanced canopy surface area per unit land area (Nalini *et al.,* 1999; Iyer and Kurian, 2006).

Canopy management and reproductive manipulation vary according to climatic conditions, cultivar and available technology (Davenport, 1993). Pruning, being a proven technology, is an integral part of canopy architecture management, which helps effective foliar applications, fertigation, integrated nutrients and pest management to achieve higher yield and quality in various perennial fruit crops (Mishra and Kaushal, 2013). The litchi plant has the tendency to bear fruits on outer surface of canopy receiving maximum solar light and economical yield is confined to two-thirds of the lower middle canopy (Singh *et al*., 2011). Nath *et al.* (2003) clearly indicated that stereo fruiting in litchi with quality fruits can be obtained by opening of centre portion of plant and facilitating more light penetration inside the tree canopy which indicates that once litchi trees begin to bear, fruit production becomes simply a function of effective canopy surface area of outer and inner side of the crown structure.

There is a linear relationship between total leaf area and canopy surface area passing through the origin (Menzel and Waite, 2005). Fruit development in litchi is also sensitive to light conditions (Yuan and Huang, 1988). Therefore, being a very crucial component of productivity vis-à-vis the all management practices, crown structure and canopy size in litchi is of paramount significance. The nutrient application is also very vital in litchi for canopy development, leading to the sustainable production and productivity. Keeping this in view, many attempts have been made to see the combined effect of pruning practices and nutrient application to create a congenial / healthy canopy environment for quality fruit production which resulted significant outcome (Kalloo *et al*., 2005; Lal *et al*., 2000; Kumar, 2008; Singh and Mishra, 2007; Singh *et al.* 2011). There is occurrence of dominance of vegetative phase over reproductive phase in these perennial evergreen fruit trees. Evergreen trees bound to grow continuously due to abundant sunlight throughout the year. Dominant vegetative phase can affect the reproductive phase especially under tropical regions, if not regulated properly (Srivastava, 2007).

Canopy management has direct correlation with dry matter production, flower bud initiation and fruit quality. The canopy physically supports the fruit crop and markedly influences the yield and quality of fruits. Innovations in production system have witnessed significant variations in tree canopy forms. Huang *et al.* (1992) studied the relative productivity of old orchard (in Fujian) with different sized canopies and showed that yield was strongly correlated with the diameter of the tree crown and concluded that best yields were associated with trees that had two to four main branches and a trunk girth of at least 1.0m. Canopy re-building by proper and corrective cutting, pruning and training system optimize the production of assimilation and its conversion into economic yield. Through knowledge for understanding of canopy development and management is of paramount importance to achieve optimum efficiency of orchard after rejuvenating the declining unproductive trees. Canopy geometry, modification pruning, training and application of growth regulators to control vegetative growth are designed to improve light penetration, thereby increasing the overall fruit bearing potential and improving yield and fruit quality. Therefore in order to optimize the utilization of light for increased yield of quality fruits, canopy management deserves greater attention by exploiting the various available techniques, i.e. through the selection of proper rootstock and scion combinations, training and pruning systems, cultural practices, plant growth regulators, suitable nutrient and water management practices and tree and branch orientation would help in maintaining the ideal canopies of trees.

Training systems have drawn considerable attention over the past 40 years since they must combine different purposes. Those have been revealed to be more or less conflicting, since economic conditions have varied over time. The main purposes, especially in intensive orchards, are : (1) a rapid achievement of a developed canopy structure to reach orchard maturity and maximum fruit production within a few years; (2) an optimal capture of light to optimize carbon gain and fruit yield per hectare; (3) a fair distribution of intercepted light within the aerial system of tree to minimize the spatial heterogeneity of local vegetative growth and fruit quality; and (4) management of tree shape and fruit load with minimal pruning, to take advantage of natural trends of cultivar and reduce the economic cost of this manual operation.

The last point is of major importance since training systems initially conceived to improve light interception by the tree overall may stimulate growth of vigorous water-shoots, i.e. reiterated complexes, on the upper side of scaffold branches. If not removed, these shoots acting mainly as assimilate sinks may also thwart the benefits of high illumination within the tree by decreasing light interception by fruiting shoots. On the other hand, an unpruned tree, in which vigour is well-distributed to fruiting shoots, quickly begins production but, in most cases, results in an overcrowded canopy after some years and eventually fruit size and quality are reduced. Training methods have then been particularly developed at the tree scale to manipulate both the vegetative and the fruiting components. Pruning vegetative shoots at different positions in the tree or/and at different phenological stages is used both for the building of the tree structure, according to a specific tree shape, and to optimize light distribution within the canopy as in cherry (Flore *et al.*, 1996) or in apple (Barritt, 1992). One is to maintain the tree in the allotted space in relation with the tree management system. It is preferred to heading cuts for the control of tree growth and shape by proper training procedures.

Citrus trees are self shape forming. Growers should select the correct time of their pruning because citrus trees are evergreens, they do not have a period of true dormancy but less active period is just after harvesting. Pruning to create a scaffolding network helps assure maximum fruit yield. Pruning for removal of water sprouts from young and old tree is important (Srivastava, 2007). In cashew, new flushes arise from the dormant buds on the trunk. These new sprouts form the canopy within a period of at least 10-11 months depending on variety and soil type (Kalloo *et al.*, 2005). It has been ascertained that canopy management through advanced training systems in fruit plants has been found more effective. The canopy model also influences the water-use efficiency. The objective of canopy management lies in the fact, as to how best we manipulate the tree vigour and use of maximum available light and temperature to increase productivity and quality, and minimize the adverse effect of weather. The approach requires a high degree of horticultural skill to manage the canopy of the tree.

Irrigation

Irrigation timings in perennial fruit crops (mango and litchi) is very important for its effective utilization and it should be applied before attaining the stress. The physical factors in the method of irrigation or its system selection include soils, climate, topography, water quality and availability, water table depth, orchard size, system performance, maintenance and repair etc., while irrigation practices are influenced by available technology and cost, soil type and depth, rainfall amount and distribution, fertilizer practices and production objectives (Crane *et al.,* 2009; Menzel and Waite, 2005; Singh *et al.*, 2012). Rejuvenated trees require immediate irrigation if any dry spell occurs just after the reiterative pruning operation is over, as to avoid drying out of the trees. Irrigation is must after the manure and fertilizer application. Preliminary research suggest that irrigation at 7-day intervals during the period of fruit development in mango increases fruit size, earliness and yield (Larson *et al.,* 1989). In principle, optimum soil moisture should be maintained after the emergence of new shoots on pruned trees. Almost in all perennial fruit crops except aonla, depending upon the temperature, soil type and soil moisture status, pruned trees must be irrigated at regular interval of 08-10 days during hot summer season and 15-16 days during winter season helps fast vegetative growth and good canopy development. Irrigation applied to the intercrops also gives the added advantage to the rejuvenated trees. In aonla and pomegranate, light irrigation just after manure and fertilizer application, as this fruit crop is highly susceptible to water logging condition, while irrigation is required in summer approximately at fortnightly intervals, while in case of citrus no irrigation / water logging condition near trunk is suggested (Maity and Das, 2004). Among various irrigation systems, basin system is well suited but drip irrigation has also given better response.

Irrigation to maintain the enough moisture content for two years is very crucial. The irrigation schedule should be started just after one month of pruning operation along with nutrient application. During the third year, application of water through irrigation is restricted/withheld from November and onwards till February end of the next year (mainly during winter months). Soil water content influences the frequency of flowering, the number, length and panicles, yield and quality of perennial fruits (Chang and Lu, 1995). Till date, irrigation practices are based on observation and experience (Crane *et al.*, 2009).

Mulching

Mulching is a practice of covering surface of soil with organic and polythene to check evaporative loss, modify soil thermal regime, check weed population and their growth and improve the availability of water and nutrients to the plants. It is advisable to do mulching in the basins around the pruned tree trunk for conserving soil moisture in the basins during April-June and November-January. Various mulches in rejuvenated plants in the orchards have been found better with paddy straw, sugarcane thrash, dried grasses, mango/litchi/banana dried leaves etc applied thick enough (10-15cm) to prevent weed growth simultaneously not stopping rain water penetration to the root area or rhizosphere. Black polythene (200/400-gauge) sheet might also be used for mulching in most befitting manner. As of late, plastic films have come into use for purpose of mulching due to its inherent advantage of efficient moisture conservation, weed control and maintaining of soil temperature. Besides, the volume of plastics material to be used per unit area is much less or than that of traditional mulch material and, hence, plastic films is easy to handle, transport and lay in the field and it has been successfully experimented in rejuvenated plots of many perennial tropical and subtropical fruit crops (Singh and Mishra, 2007; Singh, 2005; Janik and Paull, 2008; Singh, 2010). Mulching at the base immediately surrounding the main trunk of the pruned trees during April-May and September-October have been found beneficial in conserving soil moisture beneath the tree canopy, reducing the frequency of irrigation and enhancing growth rate of vegetative flush. In citrus, organic mulches in sweet orange, dry leaf mulching in Coorg mandarin and sod mulch in mosambi orchard have found influencing soil tilth and enhancing quality fruit yield (Maity and Das, 2004).

Intercropping

During initial 2-3 years after rejuvenation, there is excellent opportunity for economic utilization of inter-space in rejuvenated orchards. The open interspaces between the plants/trees is created in orchard like newly-planted orchard. Intercrops during summer season and during rabi season have been found most suitable. Apart from the significant income obtained by intercrops in rejuvenated orchards, added advantages like improvement in soil tilth (health), almost complete check over weed population/growth as well as less incidence of pests – diseases are also received. In general, if rejuvenated plants / tress attain full growth again and start giving commercial viable yield, intercropping may be discontinued. Even though, some of the important principles should be observed, particularly for growing of intercrops in the open space created in fruit orchards after reiterative pruning, are: (1) intercrops should not occupy or disturb/interfere the area where the roots of fruit trees are concentrated; (2) soil fertility should be maintained or improved when intercrops are grown; (3) water requirements of the intercrops should not clash with those of the main fruit trees (intercrop may require an irrigation at a time when it would be detrimental to rejuvenated trees); (4) intercrops should be selected with reference to their effect on soil moisture; (5) grain crops should be avoided as they remove excessive moisture to the detriment of fruit trees and the intercrops selected should not exhaust the soil water and nutrients and should not demand more water than is allowed for fruit trees. The most important aspect is that intercropping should be started at least after three months, when reiterative pruning operation is over. Intercropping with suitable crop and crops sequences have been found very encouraging and beneficial.

Nutritional and Physiological Disorders

While rejuvenating the perennial fruit trees, the invigoration effect after hard core pruning is instantly managed by the stored food material and the existing root system of the rhizosphere. Any imbalance and ill management after pruning operation is over and that too during regeneration growth and development process, may be fatal if any deficiency/excess of nutrient elements, any hormonal imbalance or abiotic stresses (adverse climatic or growing conditions) occur.

These may cause physiological disorders due to one factor or may be the combination of so many interdependent factors, in rejuvenated fruit trees, which are vulnerable to different types of physiological disorders, which may cause tree mortality, leading to huge loss to the growers (Sharma, 2006). Hence after hard core reiterative pruning, one should be careful about the important physiological disorders of that particular perennial fruit crops, which required timely management practices. In general, climate influences growth and productivity of fruit trees, much more than any other factor which affect the fruit plants, hence for rejuvenation, careful management and input use efficiency is be taken care with utmost priority for fast regeneration, vegetative growth and canopy developmental processes (Kumar, 2008).

Plant Protection

Successful pest management programmes have been developed in different high-value fruit crops. The principles of fruit pest control have led to the evaluation of damage to express population trend and extent of damage. In principle, disease management is to reduce the inoculum density in nature and to restrict the spread of inoculum for infestation, which has been attributed to cultural method, regulatory method, chemotherapy and biopesticide control. By controlling the insect and pests, restoration of vigour and keeping the tree more healthy, help in enhanced and sustained quality fruit production (Kalloo *et al.*, 2005; Kumar, 2008; Singh, 2010; Singh and Mishra, 2007). Intensive care to control the infestation of important pests and the diseases in tropical and subtropical perennial fruit crops are required (Dodd *et al.*, 1997; Masarrat *et al.*, 2000; Ploetz and Prakash, 1997; Ploetz, 2003). Control methods may be biological, mechanical or chemical or the combination depending upon the growers' preference and specific circumstances. Leaf roller, leaf miner and leaf cutting weevil starts damaging right from the beginning of the rejuvenated trees, hence care is required from that moment itself, as these are attracted to new vegetative growths. The proper IPM practices are also required to be taken care off for further sustainable and quality fruit production

Fruit Yield and Quality

Fruit yield and physicochemical characters of mature fruits are found to be better in fruits obtained from rejuvenated trees. Maturity period is found to be slightly delayed in rejuvenated plants. Fruit yield in second year in guava, citrus and aonla, while third year in rejuvenated trees of mango, litchi and cashew obtained with enhanced quality production. The categorization of fruit under different quality class revealed high percentage under extra class and class I as compared to un-pruned trees. The quantum of fruit yield with enhanced quality increased significantly every year onwards after the starting of bearing in rejuvenated trees.

Economics

Various operations in rejuvenating old senile trees are not an immediate process, but usually they can be done effectively over a period of 2-3 years and even later maintenance and management continues. Economics of rejuvenation of old senile orchards in some of the perennial fruit crops are required to be calculated for 3-5 consecutive years with particular operations like sale of fruit before rejuvenation, cost of reiterative pruning, sale of woods, cost of intercropping, profits from intercrops, sale of fruits from rejuvenated trees and finally gross income to compare with benefit: cost ratio, and other added advantages. The experimental attempts of pruning proved to be successful in rejuvenating old orchards and the economics also suggested gain in returns compared to senile orchards (Singh *et al.,* 2012). Pruning schedule for rejuvenation in different citrus species played a key role in improving fruit yield considerably (Ansari *et al*., 2011). In aonla, all the old senile trees responded well to rejuvenation and the technology helped in securing quality yield (Singh and Mishra, 2007). In cashew, yield levels increase at significant level. The result of rejuvenation in pomegranate showed that fruit yield/ plant found surpassing control even in second year itself and the increase was in progressive manner (Hiwale, 2009).

Hitch in Adoption and Strategic Need

The reports of survey and investigations made by the institutes/SAUs and Government departments have clearly indicated that there exists a wide gap between technology availability and technology adoption. High commercial yield with quality production depends on high level of technological convergence and adoption (Kumar, 2012). Depending on need and profitability, farmers / orchardists generally integrate new technologies with their ongoing management and cultural practices. There is not any hitch in adoption of safe, secured, non-invasive, short-term, efficient and profitable technologies and these type of efficient and economic technologies are giving good impact within short gap only. It is difficult to achieve the desired outcome through the use of single technology centric approach or the use of component technology in isolation. The rejuvenation technology is having sound blending of the series of measures and practices developed for integration of reiterative pruning, followed by nutritional, irrigation, mulching, thinning of shoots and training for canopy rebuilding in desired shape, plant protection along with intercropping in open space created after pruning for a period of three years without any economic and existing plantation loss.

The rejuvenation technology is a medium-term strategic programme for old senile, woody perennial fruit crops, with sound blending of series of practices developed for making complete transformation from their unproductive old senile phase to young productive economical phase in sustainable manner (Kumar 2008, 2013; Lal *et al.,* 2000). The technology envisages to exploit potential of the productive (commercial) phase of the perennial fruit crops to generate maximum output, profit, quality along with resource use efficiency with maximum environmental protection. Minimizing yield gaps, increasing profits and quality fruit production are need of the hour to ensure food and environmental security.

In spite of vital, viable and economically-feasible technology, the invasive nature of operation of the technology posing major hindrances in the speedy adoption of rejuvenation technology. The only hitch at the initial stage is lack of trained people and required proper tools / equipment with operator to complete the operation successfully. The efforts have been made to organize several onsite interactions and practical demonstration trainings by state departments, SAUs and research institutes, which has successfully created enough awareness and tendency to go for this technology for their old senile unproductive fruit orchards. The only hitch in adoption is that lack of team of skilled persons ready to be available on demand to translate this invasive technology in efficient and effective manner. The lack of skilled work force and poor technology transfer mechanism in this regard invites the urgent need for strong technology implementation or transfer system (Kumar, 2013). The efforts are required to constitute team of personnel (trained and expert hands) for skillful and successful operation of technology in coordinated way on demand and payment basis in an economically-viable manner. The central and state level institutions and organizations are now involved for monitoring the technology by providing training with financial assistance as well as subsidy plan (National Horticulture Board, National/State Horticulture Mission, ICAR Institutes, State Agricultural Universities and Krishi Vigyan Kendras and even some NGOs). The hindrances in significant adoption is lack of implementing team equipped with required machineries (Singh *et al.,* 2011; Kumar, 2012). The Hi-tech tools and machines required for this technology with expert operators should be provided with the team for effective and efficient technology implementation and operation with beneficiary's utmost satisfaction and will for fast pace of technological adoption (Kumar, 2012). This type of strategic plan and efforts take care of psychological factors of the beneficiaries in convincing way and avoiding their stress process during the technological operation (Bongers and Winter, 1992; Karasek and Theorell, 1990).

Impact

With the scientific skill and approach, complete transformation have been brought about by changing the exhausted phase of the old senile perennial fruit crop orchards to behave and bear like young commercial orchard with sustained performance. This transformation gave great impetus and boost to the enhanced quality fruit production having high economic returns. This wonder became possible only with the adoption of scientific approach. This technology has shown the path for enhancing the production and quality of fruits year after year without much of environmental stress and cost involvement. It is of great importance for rejuvenation of the aged and declining perennial fruit trees and improvement of the fruit production management to enhance output and quality of fruits. It is indeed became an example, when the utmost transformation tried through rejuvenation pruning of old senile, unproductive orchard, without destroying the actual identity and fortunately, only in time gap of two to three years it started behaving like young commercial orchard with sustained economic returns. As a result, the increasing trend of production and productivity of quality yield in coming years as commercial bearing orchards, the fruit growers having old plantations giving uneconomical yield should adopt the technology developed scientifically by various National Research Institutes/State Agricultural Universities/Government Departments. Region-specific various on-site training programmes and demonstration trials have been conducted and still going on to increase the level of adoption by the actual beneficiaries (Kumar, 2008, Singh *et al.,* 2011). Realizing the need and importance of rejuvenation technology, the National Horticulture Mission (NHM) of different states have included the particular technology package in its action plan. The National Horticulture Board (NHB), Gurgaon, Haryana has funded projects and National Banking for Agriculture and Rural Development (NBARD) has already initiated programmes with enough subsidy to take technology to farmers field. The mass communication media particularly local newspaper published and flashed and even regional TV channels have made many telecasts this successful technology transfer with various success stories. Overall, the success of rejuvenation technology in transforming the old senile orchards into young commercial bearing orchards have become the most important consideration in National Policies of Government Horticulture Developmental Plan in almost all the major perennial fruit growing states.

Chapter 7

Aonla

Aonla or Indian gooseberry (*Emblica officinalis* Gaertn), belongs to family Euphorbiaceae, is indigenous to Indian subcontinents (Morton, 1960). It has become an important commercial fruit crop in India, due to its hardy nature, suitability to various waste lands, high productivity, nutritive and therapeutic values (Singh *et al.*, 2003). In its processed form, is very popular among the social elites (Singh, 2012). As an indigenous fruit, it has extensive adaptability to grow in diverse climatic and soil conditions ranging from plain lands to hills. The climate ranging from hot tropical plains to humid subtropical mid-elevation hills has been found suitable for its cultivation. It is successfully raised in arid, semi-arid, coastal and warm temperate conditions. Similarly it grows well in saline, alkaline and degraded as well as in sandy, red and clay soils (Singh, 2012).

It is not merely a source of nutrients and medicines but its cultivation is also highly remunerative for small and marginal farmers. Traditionally, aonla has been a common crop of forest or household, but during the last decade or two, unprecedented expansion has been witnessed in the sphere of aonla cultivation across the country, utilizing the wasteland. Many small and big orchards have been established. This has resulted in efficient utilization or resources culminating in increase in farmers' income and nutritional security coupled with enhanced employment and better utilization of wastelands.

PRESENT STATUS

Aonla is a wonder fruit of new commercial interest in India and our country ranks first in the world in area and production. Apart from India, naturally-growing aonla trees are also found in Sri Lanka, Cuba, Puerto-Rico, USA (Hawaii and Florida), Iran, Iraq, Pakistan, China, Malaysia, Bhutan, Thailand, Vietnam, Philippines, Trinidad, Panama and Japan. In India, major aonla growing states in India are UP, Maharashtra, Gujarat, Rajsthan, Andhra Pradesh, Tamil Nadu, Karnataka, Haryna, Punjab and Himachal Pradesh (Singh, 2012).

Uttar Pradesh ranks first in area and production. The major aonla growing areas in Uttar Pradesh are Pratapgarh, Raibareli, Varanasi, Jaunpur, Sultanpur, Kanpur, Fatehpur, Agra and Mathura districts. In Madhya Pradesh, major aonla producing regions are Dewas, Hoshangabad, Shiwani, Tikamgarh, Betul, Chindwara, Shivapurkala, Panna, Rewa and Satna. In Haryana, aonla is mainly grown in Bewal and Gurugaon areas. In Karnataka, Bilgiri Rangan Hills in Mysore is aonla growing area. In Tamil Nadu, concentrated production is around Salem and Dindugal. In Himachal Pradesh, aonla is grown in Palampur, Bilaspur and Hamirpur. Uttar Pradesh is the largest producer of cultivated aonla, with an annual production of 63,000 tonnes from 15,700 ha with an average productivity of 4.0t/ha (Singh and Mishra, 2007). The second and third largest state producing aonla is Gujarat and Tamil Nadu. The highest productivity of aonla (5.2 t/ha) is from Haryana, whereas lowest from Rajsthan (1.2 t/ha).

PRODUCTION CONSTRAINTS

Since most of the aonla orchards in India are of seedling nature, they are poor yielding with inferior quality fruits. There is very low level of adoption of standardized efficient and economically-viable cropping system. There is also lack of knowledge and effective adoption for plant protection measures like IPM, use of bioagents, biopesticides and biofungicides even at the commercial bearing stage, leads to decline of the trees (Mehta and Srivastava 2008). One of the reasons for low productivity is a large number of aonla orchards in the age group of 30- to 50 years and above, have either gone unproductive or show marked decline in productivity. Most of the old plantation of aonla fruit belts are declining and showing poor productivity and poor quality of fruits. The sole cause of declination in production and quality is due to poor orchard management and overage trees. This attributed to over-crowded and intermingling of large branches and meagre foliage, allowing poor light availability to growing to growing shoots within the canopy. This renders them uneconomical. Other constraints are as branches of aonla trees often break off carrying heavy crop load due to the brittle nature of wood. Water sprouts and rootstock growth poses a problem under favourable environmental conditions. Thus, there is dire need to improve the health and production of senile orchards through application of rejuvenation technology. The practice of pruning supplemented with balanced doses of nutrients is not properly followed. Although, it is highly remunerative crop, yet the productivity and quality of fruits of senile orchards are being declining due to poor management practices as most of the cultivators do not apply manures and fertilizers that resulted in diminishing the fruit size and quality.

Yet the old plantations of many varieties possess shy bearing tendency and produce poor fruit with inferior quality (Tiwari *et al.*, 2015).

Need of rejuvenation

Till date, there is considerable area under aonla orchards in India are of seedling nature, giving very poor fruit yield or have become unproductive. These are required to be transformed into productive ones with superior varieties. In case of aonla commercial orchards there are many disadvantages with large and crowded trees (Kalloo *et al.*, 2005; Singh and Mishra, 2007). In aonla, there is a natural progression from a single shoot tree towards tall. These disadvantages have resulted in widespread efforts to reduce tree size, accelerate canopy and yield development and improve canopy form to overcome the limitations of the large tree. The most important measures of orchard efficiency are total fruit yield or yield of high quality fruit per unit of input (Land, capital, labour, light, energy, water and nutrients). The fruit growers are also interested in efficiency expressed in terms of managing the orchard canopy through rejuvenation and top working. As pruning is very effective to induce healthy current season shoots from older branches, causing invigorating effect, sprouting to more number of current season shoots, increased level of available soil nutrients and decrease in utilization of excessive nutrients by undesirable branches, which ultimately result in obtained high quality fruit yield in sustained manner (Tiwari *et al.*, 2015).

Identification of old senile/unproductive orchards

Most of the aonla orchards in India are of seedling nature, which have grown older and exhausted can be selected. The orchards which remained neglected and attained the declining stage having poor lanky growth, highly infested with disease and pests, should be targeted for implementation of rejuvenation (Singh and Mishra, 2007).

Training to growers

Exhausted aonla trees can be rejuvenated by heading back of branches in winter for the production of new shoots, which can bear good crop in the following years. A technology to rejuvenate aonla trees and enhance the production potential of senile orchard has been developed which involves rejuvenation by drastic pruning of branches, followed by management of shoot as well as by top working (Kalloo *et al.*, 2005; Singh and Mihsra, 2007).

Most of the aonla orchards in India are of seedling nature and need to be rejuvenated by top working. Fruits obtained from these seedling tees are generally smaller and inferior in terms of quality. Such established and healthy orchards can be converted into productive ones by top working with improved and commercial cultivars.

There was a loss of two fruiting seasons from rejuvenated trees as compared to control. From commercial point of view, the fruit quality of control trees is inferior with smaller fruit size. In rejuvenated trees, fruiting starts from third year onwards as meagre fruiting with improved fruit quality of commercial importance. There was a lot of improvement in fruiting of rejuvenated trees in fourth year with maximum fruiting in NA-7 cultivar. Meagre fruiting in rejuvenated trees during first bearing is due to comparatively smaller canopy area than the seedling trees. The quality of fruits obtained from top worked trees was significantly better than fruits obtained from seedling trees. Fruit length and fruit width were significantly higher in fruits obtained from top worked trees. Similarly fruit length and fruit breadth was recorded least for fruits obtained from seedling trees.

REJUVENATION TECHNIQUE

Old and unproductive trees of aonla can be rejuvenated to make productive and easily changed into superior type by top working (Singh, 1953; Singh, 2012). The plants are headed back during December-January 2.5-3.0 m above ground level. The cut portion of the trees is then pasted with the tree paste (cowdung + soil+water) or simple cowdung paste to avoid infection of fungal diseases. Several new shoots are induced on pruned branches after heading back and a few shoots are retained at proper spacing and growing towards periphery of trees. Successive removal of unwanted shoot, considering the vigour and growing direction is important. In this technique only 4 to 6 shoots developing in outer direction on main limbs should be allowed to develop. Proper development of new canopy in horizontal direction should be kept in mind while practicing thinning of shoots. So, plants should be trained to modified central leader system. Two to four branches with wide crotch angle, appearing in the opposite directions should be encouraged in early years. During March – April, prune and thin the crowded branches to provide maximum fruit bearing area in the tree. During June-July, the patch budding with superior variety is done on each selected shoot. After sprouting, the top portion of the shoot removed. Care should be taken to manage the insect –pest problems during this period, since these plants are prone to insect and sometimes to wind damage.

Top-working

For budding operation, the scion shoot should be selected from the mother plants, which are prolific bearer and free from disease and pest incidence. The scion shoots of superior variety can be budded on the newly emerged shoots

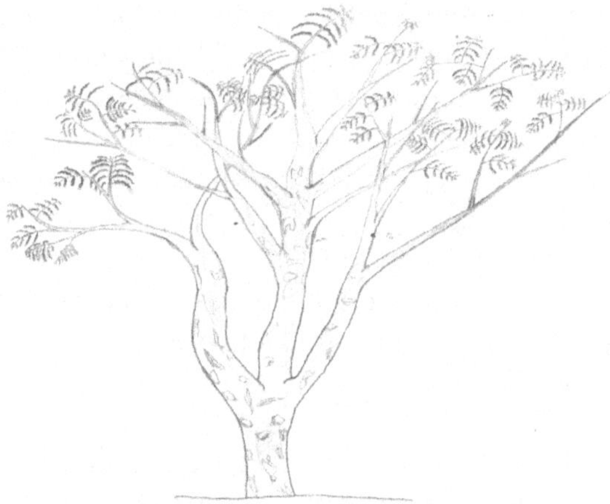

Old senile unproductive aonla tree

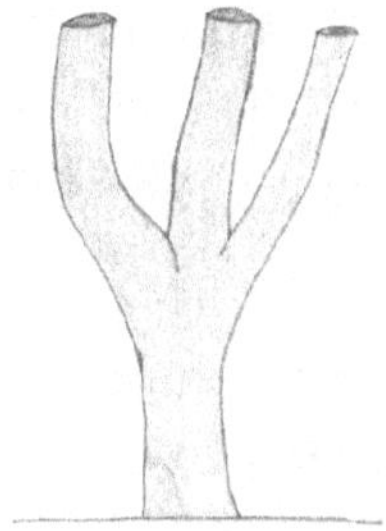

Pruned tree

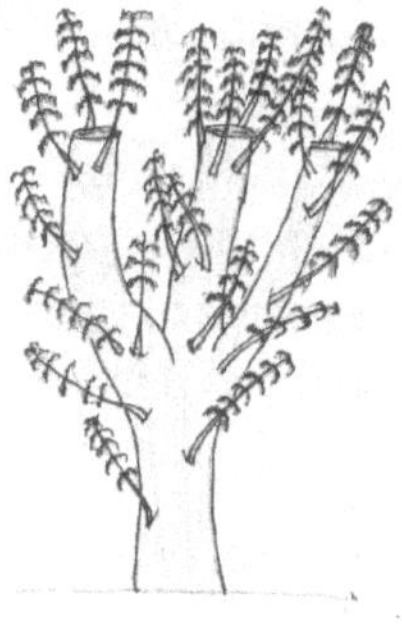

Emergence of shoots

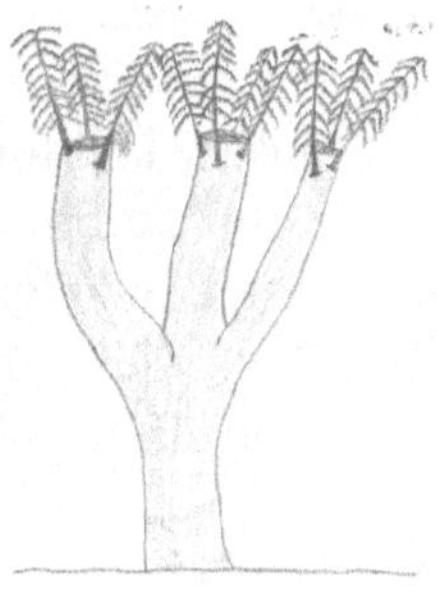

Thinning of shoots

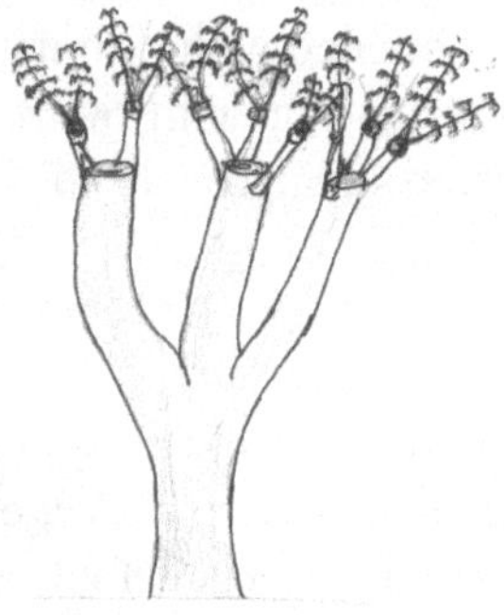

Top grafting

Rejuenated tree in production

developed on beheaded branches by adopting patchs/modified ring budding from May to September. After successfully top-worked side shoots which emerge on the pruned branches, should be removed regularly as and when they emerge. From third year of budding, the plants grow into tree and give flowers and fruits like a normal tree. Top worked on rejuvenated trees initiated flowering and fruiting from third year onwards (Mishra *et al.*, 2007). During the first year the production is less due to smaller canopy area and less number of hermaphrodite flowers on new growth. The yield continues to increase in succeeding year, turning the unproductive tree into productive ones.

Nutrition

During the phase of rejuvenation in aonla a dose of 50 kg well rotten FYM, 1.50 kg Urea, 2.00 kg Single Super Phosphate and 1.0 kg Muriate of Potash should be applied after one month of rejuvenation preferably in two split doses one during the February-March and remaining half dose in August (Singh and Mishra 2007), while another recommendation manuring may be done in July-August and second in December January (Maity and Das 2004) and in both the time manure and fertilizer should be spread around the main trunk and mixed well in the surface soil followed by light irrigation. In sodic land, 100g each of borax, zinc sulphate and copper sulphate should be applied along with fertilizers as per tree age and vigour. Besides this, spray of micronutrients, viz. B, Zn, Cu (0.04%) along with hydrated lime is also done. It is helpful in reducing fruit drop and improving fruit quality.

Irrigation

Normally, irrigation is not much required in aonla plantation. Aonla plantation is highly susceptible to water logging condition. However, adequate irrigation is required during heading back of branches for better canopy development. After heading back of branches irrigation at an interval of 7 to 10 days is desirable for initiation of new shoots. Among the various irrigation systems basin system is well suited for aonla. The initial information obtained with drip irrigation has shown promising response. Alternate drip irrigation with 60 % open pan evaporation (OPE) is appropriate with a water saving of 40 to 45 per cent (Singh, 2012).

Intercropping

During initial 2-3 years after rejuvenation, there is excellent opportunity for economic utilization of interspace in the rejuvenated orchard. Vegetables like bottle-gourd, okra, cauliflower, coriander, turmeric and elephant foot yam, flowers like gladiolus and marigold have been found well suited for intercropping in rejuvenated aonla orchards. In salt affected soil or marginal soils, intercropping of *Dhaincha* for two years is beneficial for improving the physico-chemical properties of the soil.

Mulching

Mulching is a practice of covering the surface of soil with organic and polythene to check evaporative loss, modify soil thermal regime and improve the availability of water and nutrients to plants. Various mulches in rejuvenated aonla orchards are found better with paddy straw, sugarcane trash, coconut husk and FYM. Apart from increasing growth and yield, it also improves organic content, infiltration rate, restricting the upward movement of soluble salts and thus escaping their toxicity menace in salt affected soil. Among various mulching materials, black polythene (100 micron) and paddy straw proved best materials in respect of yield, number of fruits, nutrient status, water use efficiency and soil microbial population. Mulching with organic wastes at the base upto 15 cm height during summer, not only helped to conserve moisture but also helped in keeping weeds under check (Mehta and Srivastava, 2008; Singh, 2012).

Management of Physiological Disorders

Internal necrosis : Internal necrosis has been observed in aonla fruits. Aonla cultivar, Francis is highly susceptible followed by Banarsi. The symptoms start with browning of inner most part of mesocarpic tissue at the time of endocarp hardening in the second or third week of September which later extends towards the epicarp resulting into brownish black appearance of flesh (Singh *et al.*, 2013). Infection is noticed on other aonla cultivars like Chakiya, NA-6 and NA-7. Internal fruit necrosis is associated with boron deficiency during the fruit development (Singh, 2012).

Cultivation of cultivars Chakia, NA-6 and NA-7 need to be encouraged. Combined spray of Zinc sulphate (0.4%) + copper sulphate (0.4%) and borax (0.4%) during September – October has been found effective.

Fruit drop : It is a serious problem in aonla which influences the final yield of crop. There are three waves of flower and fruit drop. In first wave, more than 70 per cent flowers drop within three weeks of flowering which is normally due to lack of pollination; in second wave, second drop is that of young fruitlets at the time of dormancy break, and in the third wave, third drop of fruits is due to embryological and physiological factors and it is spread over the entire period of the fruit development. It starts from later half of August and continues up to harvest (Singh *et al.*, 2013).

Fruit drop results due to many factors such as : 1, dry spell; 2, imbalance of hormones; 3, improper nutrition; 4, fluctuation in temperature; 5, variety and age of tree; and 6, number of developing fruits. The delay in harvesting results in heavy fruit drop particularly in Banarasi and Francis cultivars.

Management of Pest

Shoot gall maker (*Betousa stylophora* S.) : The pest attacks nursery plants and old bearing trees. The insect is active from June to December. The pest causes gall formation on stems and shoots. Its larva of this moth tunnels in the apical portion of the shoot and infested for bulge into gall. This pest is important as its large scale infestation may result in stunted growth of the trees, affecting flowering and fruiting. In the beginning of the infestation terminal shoots swell, which increase in size with the passage of time. Full sized galls can be seed in the month of October-November.

Over crowding of branches should be discouraged. Galled shoots should be pruned and destroyed along with the pest after harvest. Sprays of 0.05% monocrotophos in beginning of the season should be done.

Bark eating caterpillar (*Indarbela tetraonis* Moore): This pest is widely distributed through out the country. Generally, the incidence of this pest is more common in neglected and poorly managed aonla orchards. The infestation of this pest starts in the month of April with the emergence of moths. The attack of this pest may be identified by the presence of irregular tunnels and patches covered with silken web consisting of excreta and chewed up wood particles, on shoots, branches and trunk (Singh *et al.*, 2013).

Kill the caterpillers mechanically by inserting the iron spike in shelter holes made by these borers at early stage of infestation or soon after heading back of branches. In case of severe infestation remove the webs and insert the swab of cotton wool soaked in 0.025% dichlorovos or inject water emulsion of 0.05% monocrotophos or 0.05% chlorpyriphos and plug the holes with mud.

Mealy bug (*Nipaecoccus viridis*) Incidence of mealy bugs has been noticed from March to July with the peak population in April May. Excessive desaping by nymphs results in drying and dropping of leaves and flowers which affect growth of the tree, flowering and fruiting (Singh *et al.*, 2013).

Prune affected parts and destroy them at early stage of infestation. In case of severe infestation spray 0.05% monocrotophos or 0.05% quinalphos.

Aonla aphid (*Cerciaphis embelica*) : The incidence of this pest is mainly seen from July to October with the peak period in September. The new shoots are infested at growing points. The nymphs and adult females suck the sap.

The leaves turn yellow and dry up. Presence of ants also indicate the infestation of aphids (Singh *et al.*, 2013).

Clipping off and destruction of affected leaf and shoot should be done. The aphids should be managed by spraying of 0.06 dimethoate or 0.05% monocrotophos.

Stone borer (*Curculio* sp.) : The pest is found active from June to January. Tiny weevil emerges in the month of June with the onset of rains. The eggs are laid by excavating a cavity below the epicarp. Fruits of 1.5 to 2.0 cm in diameter are preferred for oviposition. Full grown larva comes out of the fruit by cutting small circular hole.

Deep ploughing of the orchards after harvesting exposes the diapausing larvae and is effective in bringing down the pest population. First spray of 0.2% carbaryl or 0.04% monocrotophos or 0.05 % quinalphos needs to be done at pea size of fruit. Second spray may be done at fortnightly interval with changed insecticide, if needed.

Fruit midge (*Clinodiplosis* sp.) : Incidence of this pest occurs in the fruiting season in aonla from September to January. The eggs are laid inside the fruits. Immature larvae are creamish white, while mature are pinkish – orange in colour. In the beginning, small grey black spots appear at the site of infestation. Desi varieties of aonla are more susceptible.

Spray of 0.2% carbaryl or 0.04% monocrotophos or 0.05% quinalphos at the beginning of the fruiting is recommended.

Management of Disease

There are few important diseases, which are very dangerous for aonla cultivation. As fruits are perishable in nature, even slight injury causes diseases in storage and transit. Some of the diseases of aonla are described as follows :

Rust (*Ravelnelia embelica* Styd.) : Rust of aonla is an important problem specially in Rajasthan and has been reported very serious problem in Udaipur district. Recently, it has been recorded in desi varieties of aonla in areas of Lucknow and Pratapgarh. On fruits initially, few black pustules appear which later develop in a ring. The pustules join together and cover big area of the fruit. The black spores are exposed after rupturing a papery covering. Fruits give dirty look and lose its market value. On leaves pinkish brown pustules develop which may be arranged in groups or scattered as isolated pustule (Tyagi and Pathak, 1987; Singh *et al.*, 2013).

Three sprays of wettable sulphur (0.4%) at an interval of one month from July has been found useful in controlling the disease.

Wilt : Recently, wilting of aonla plants have been observed in different parts of India. The main reason for wilting though was attributed due to frost injury but association of Fusarium sp., is also found.

Management : Heading back of branches followed by mulching with organic waste or black polythene and cow dung pasting on tree trunk have been found quite effective in managing the wild affected trees.

Sooty mould : The aonla trees which are infested by scale insect, are generally affected by sooty mould. Sooty mould causes velvety covering of black fungal growth on surface of leaves, twigs and flowers. It can be managed by spraying of starch @2%. Insecticide monocrotophos @ 0.05% and wettable sulphur @0.2% can be mixed in starch if infestation is more,

Anthracnose (*Colletotrichum* spp./*Glomerela cingulata* (Stoenm.) Saauld and Schrenk.): Anthracnose on leaves and fruits is common during August – September. Initial symptom of the disease is in the form of minute, circular brown to grey spots with yellowish margin on leaflets. On fruits the depressed lesions are formed, which later turn dark in the centre forming acervuli often arranged in rings. Consequently the fruits become shrivelled and rot.

Fruit rot (*Alternaria alternate* (Fr.) Keissler) : The rot starts as a small brownish spherical necrotic spot which increases in circular fashion with the development of the disease. In advance stage, the spots become dark brown to black and neighbouring spots coalesce (Ram *et al.*, 1977; Singh *et al.*, 2013).

On spraying of carbendazim 0.1% should be done 15 days prior to fruit harvest. Harvesting should be done very carefully to avoid any injury to the fruits. Treatment of fruits with borax or sodium chloride checks the fruit rots.

Control : To secure heavy fruit set and reduce fruit drop watering at biweekly intervals of mature bearing plants is necessary from April to June. Also, growth regulator sprays are found useful in controlling fruit drop.

Fruit Yield and Economics

A significant increase in yield and yield parameters after implementation of rejuvenation technology have been recorded due to invigoration effect, resulting vigorous vegetative growth and increase chlorophyll content, which together accelerates the photosynthesis that increased the supply of carbohydrates to the plants (Tiwari *et al.*, 2015). Desi aonla starts bearing fruits after second years of rejuvenation, whereas topworked (budded or grafted trees) start bearing after two years yielding approximately 100 kg fruits/year/tree. The average

Thinning of shoots

Intercropping in pruned orchard

New shoots on pruned branches

Flowering on new shoots

Flowering on old shoots

Fruiting on shoots

Heavy fruiting in rejuvenated tree

yield starts after 4th year of rejuvenation and it is about 100-110q/ha. An income of Rs. 50000-55000/ha can be obtained (Mehta and Srivastava, 2008).

Table 1. Expenditure incurred in rejuvenating senile aonla orchard (One hectare)

Operation	Amount (₹)
Charges for reiterative pruning @ ₹ 100/ tree	15,700
Cost of ploughing, leveling and cleaning	1500
Cost of manure and fertilizers	10,000
Cost of ring making and fertilizer application	7,850
Cost of irrigation	2,500
Cost of chemical pasting and spraying	2,500
Cost of thinning and other operations (including budding operation on selected shoots and other related operations)	9,000
Miscelaneous (saw blade, rope, bamboo pieces etc.)	950
Total	**50,000**

Table 2. Economics of rejuvenation of old senile orchard (one hectare) for four years on yearly basis

Operations (₹)	First Year	Second Year	Third Year	Fourth Year
Sale of fruits before rejuvenation	5,000.00	-	-	-
Cost of rejuvenation	(-) 50,000.00	-	-	-
Sale of woods	40,000.00	-	-	-
Cost of intercropping	(-) 7,500.00	(-) 7,500.00	(-) 15,000.00	(-) 6,000.00
Profit from intercrops	20,000.00	19,000.00	37,500.00	22,500.00
Sale of fruits from rejuvenated trees	-	25,000.00	50,000.00	1,00,000.00
Gross income	**7,500.00**	**36,500.00**	**72,500.00**	**116,500.00**

The study on effect of canopy and nutrient management on senile orchards of Indian gooseberry cv. Francis clearly indicated that highest net return and cost benefit ratio in the treatments receiving 50 % pruning intensity + 40 kg FYM + 100 % RDF (Tiwari *et al.*, 2015).

On the basis of hand on references (Kalloo *et al.*, 2005; Singh and Mishra, 2007; Tiwari *et al.*, 2015) and growers interaction on the basis of field experience the economics of aonla rejuvenation has been calculated with the base data and performance as follows :

The 157 plants/ha with spacing of 8m X 8m. Age of plantation 20 years. The operation cost involves cost of reiterative pruning, intercultural operations, shoot thinning, budding operations, cost of inputs, sale of pruned woods and plant protection measures etc.

Chapter 8

Bael

Bael [*Aegle marmelos* (L) Correa.], a native fruit of religious importance, is well known to the Indian people for its nutritional and medicinal values. Leaves are used as sacred offering to 'Lord Shiva'. All parts of plant viz., leaves, roots, bark, fruits, seeds etc. are used in preparation of various Ayurvedic medicines. Bael fruit is rich in protein, fat, minerals (Ca and Fe) and vitamins (riboflavin, β carotene and Vitamin C). Fruits are used to improve the digestive system and to cure stomach diseases. Fruits can be processed into various beverages and preserves (Jauhari and Singh, 1971). In spite of its Indian origin and high medicinal and nutritional values, well organized orcharding of bael is not commonly seen in the country. It is mainly found in wild and semi-wild condition throughout the India. Besides, it is also grown in the homestead gardens, backyards, religious places and farmer's fields (Misra and Uniyal, 2013).

PRESENT STATUS

Though bael is a fruit crop of subtropical areas, it has got a wider adaptability and can perform well in tropical, arid and semi-arid regions (Pandey *et al.*, 2008). Bael is also cultivated in India, Sri Lanka, Indonesia, Philippines and Fiji. Main growing states are Uttar Pradesh, Bihar, West Bengal, Rajasthan, Madhya Pradesh, Uttarakhand, Chattisgarh and Orissa. Hence, the bael cultivation can be extended to waste and underutilized lands, since it has the reputation of thriving where other fruit trees cannot survive.

PRODUCTION CONSTRAINTS

There is very few improved cultivars of bael for commercial cultivation. Some popular types named after their locality are cultivated. Bael trees are treated as sacred trees, hence no pruning and training practices are being followed, which hinders the irregular emergence of new shoots, hampers to flower induction process, flowering and fruiting in bael.

Need of rejuvenation

In bearing trees, pruning is generally not advisable as its crop bears fruits on one-year old shoots. Pruning is restricted to central opening and removal of weak, dead, diseased, dried, criss-cross and broken branches after harvesting and before commencement of new flush. Remove suckers from the rootstock time to time.

REJUVENATION TECHNIQUE

Identification of Old Senile Orchards

Participatory survey work or interaction meeting should be conducted to formulate an approach that can be adequately applied for identification of old senile unproductive trees / orchards for effective orchard rejuvenation. As there are very less number of organized bael orchards and on the other side, bael growers are least interested in adoption of package of recommended practices and even after attaining of the declining phase, they are always hesitant to go for rejuvenation. Rejuvenation activities should ultimately be coordinated with market studies and germplasm, in order to insure selection of appropriate cultivars.

Training to orchard growers

In bael, the training on rejuvenation technology should be accompanied with the knowledge sharing on cultural practices, top grafting process, pruning methods and tree reformation, fertilization strategies, irrigation requirements and plant protection measures. This program can be achieved rapidly and effectively through the development of training materials and training to prepare master trainers who demonstrate rejuvenation techniques to local lead farmers.

Pruning

Bael trees may be trained in modified central leader. Pruning is done twice in a year, once in May and other in August. Training and pruning are done during early years of planting to develop good and strong framework of scaffold branches (Misra and Uniyal, 2013). Cut the main stem at a height of 1.5-2.0 m. Pruning is limited to the removal of dead and diseased twigs/branches in May, while in August healthy leaves are pruned for sale. Pruning and training in bael is not practiced generally and hence as tree grows old attains senility and unproductiveness occurs. Hence, based on the principles of other perennial fruit crops of similar nature hard core pruning for making more productive of the old trees, heading back through reiterative pruning in reiterative manner upto the height of 1.5-2.0m can be done, which results in the formation of new shoots below the cut point. Retain 3-4 well-spaced and well-oriented new shoots (for making primary branches), which may be trained skilfully in such a way that it can contribute to a good semi-circular canopy. The primary branches become mature in 6-7 months, thereafter attaining the maturity prune these primary branches to their 50% length. This induces new shoot growth on primary branches. Retain only 2-3 secondary branches per primary. Reduction in height and spreading type of canopy facilitate harvesting of fruits and skirting of downward growing branches below the canopy area help intercultural operations for increased input use efficiency. Skirting properly also keeps the tree trunk clean, i.e. without side shoots up to 60-75 cm.

Top-working

Old and uneconomical bael trees can be turned into economical and vigorous ones by top working. In this method, trees are headed back to 1.0-1.5 m above ground level during March and new shoots emerge from the stem as in case of normal rejuvenation. Here, a few healthy shoots are retained and desired scions budded on them in June. Rest all the shoots are thinned and care should be taken, not to allow extra sprouts from below the budded portion. The sprouts emerged from scion portion are allowed to develop and trained to develop canopy. In this way inferior, old and unproductive bael trees can be transformed into superior and remunerative fruit trees (Jauhari and Singh, 1971).

Nutrition

Plant produce a number of fruits hence application of manures and fertilizers is beneficial. Though, only limited work has been done on its nutritional requirement and that too for rejuvenated trees. Apply 100kg farm yard manure, 500g N, 250g P and 500g K/plant to rejuvenated orchards in split doses. This dose is based on the requirement of a commercial bearing adult trees. Half dose of N, full dose of P and half dose of K should be given after one month of hard core pruning operation and remaining half dose of N and K should be given in the last week of August. The deficiency of nitrogen, zinc and boron is common in bael orchard. It can be corrected by soil application or foliar spray (Misra and Uniyal, 2013).

Irrigation

Fertilizer application must be followed by the irrigation immediately to maintain enough moisture content in the rhizosphere, particulalry in summer months, which is required for their rapid vegetative growth and establishment. Though in bearing trees, irrigation is not required even in dry summers, as it sheds leaves and resists hot dry summers, it is very much required and properly be applied at the time of new shoot development and leaf emergence.

Managing Physiological Disorders

Fruit cracking and fruit drop are two important physiological disorders found in bael (Pal and Misra, 2005; Misra and Uniyal, 2013). These can be managed by providing good irrigation facility, making wind breaks around the orchard and by spraying borax @ 0.1% twice at full bloom and after fruit set. The deficiency of nitrogen and zinc is common in bael orchards and can be corrected by soil application or foliar spray. In case of bael, fruit drop and cracking in fruits before ripening are main problems.

Managing Pests and Diseases

More than a dozen insect-pests are found feeding on Bael.

Leaf eating caterpillar

During new leaf emergence, is causing serious problem. Application of Thiodan @ 0.1% has been found effective. *Phyllocnistis citrella, Aonidiella aurantii* and *Papilio demoleus* are the important insects which can be easily controlled by use of insecticides. Bacterial shot hole, fruit canker and gummosis are the serious diseases. Fruit cracking is the physiological disorder in some genotypes of Bael which occurs just before ripening. Bael being a hardy crop, there is no serious insect pest and diseases as of now.

Sooty mould

This disease is frequently observed in commercial bael orchards. It can be managed by spraying of wettable sulphur+chlorpyriphos/methyl parathion+ gum acacia (0.2+0.1+0.3%).

Fruit Yield

Budded and grafted plants start fruiting 4-5 years after planting whereas, seedlings after 8-10 years of planting. Bael is climacteric fruit that can ripene, off the tree, if harvested at proper maturity stage. Maturity can be judged by the change in skull colour from dark green to yellowish green. Mature fruit should be harvested individually with 5 cm fruit stalk. A full grown (10-12 years old) budded or grafted bael tree produces on an average 150-200 fruits under good management practices. After the hard core pruning for rejuvenation the bael trees start giving fruits third year onwards and that too with high quality fruits about 50-75 fruits/tree. Later, a fully developed rejuvenated tree may produce 400-500 fruits of high quality fruits depending upon the cultivar (Misra and Uniyal, 2013). These are harvested with fruit stalk when color changes to yellowish green.

Old senile unproductive bael tree

New growth from unproductive tree trunk

New shoots on old trunk

Bael fruits on Rejuvenated tree

Chapter 9

Ber

Ber (*Zizyphus mauritiana* Lamk), is an ancient fruit crop of India. It is relished for its sweet and sour fruits. The fruits are nutritious containing high amount of vitamins like C, A and B complex. The genus *Zizyphus* belongs to the family Rhamnaceae, consisting of 40 species of tropical and subtropical regions of northern hemisphere (Yadav *et al.*, 2005). The one *Zizyphus jujube* (Chinese jujube) and the other *Zizyphus mauritiana* (Indian ber) are the two important species grown widely in temperate, tropical and subtropical regions of the world. The *ruggosa* and *nummalaria* are other important species.

PRESENT STATUS

Ber, a popular fruit, is being grown as almost commercially important fruit crop. It has better adaptability in marginal soils in arid region of subtropics. It is mainly grown in India as well as different countries in central Asia, China and Taiwan. It has been truly called as poor man‘s fruit/poor man‘s apple. Ber can suitably be grown almost in all the parts of north Indian plains in Uttar Pradesh, Haryana, Punjab and Rajasthan. Seedling ber trees are found extensively growing widely in arid and semi arid areas.

PRODUCTION CONSTRAINTS

In most of the cases, planting of ber is never attended properly and till the commercial bearing starts, the intermingling of canopy contributing branches makes the dense canopy and drastic reduction in fruiting branches. Unless ber trees are trained, these become bushy, unproductive and hinder all intercultural operations and give only poor yield. The trees also get infested with lac insect, powdery mildew etc.

Need of rejuvenation

Pruning is an essential operation to maintain vigour of trees, fruit productivity and yield of ber (Singh *et al.*, 2004). Moreover, annual pruning is required to replace old and unproductive wood by new one. In unpruned trees, the old wood goes on accumulating every year and leads to barren centre, reduced productivity and poor fruit yield. Therefore, it is essential practice to maintain their vigour and productivity as well as to improve the fruit size and fruit yield. The only option left thereafter to go for hard core pruning and again proper training of current shoots for fruit bearing and that too of quality grades (Bal, 2006).

The best time of pruning operation is the hot, dry season when plants shed their leaves and go to rest or become dormant (May). To establish a strong frame work, proper pruning and training (single stem up to a height of 75 - 90 cm in the first year, later, three to five main branches widely spread at all directions may be encouraged) is must.

REJUVENATION TECHNIQUE

Identification/Selection of Old Orchards

The ber trees if left un-pruned for a number of years may become uneconomical. These trees produce little crop of small sized fruits. In such cases the trees can be rejuvenated. Severe pruning may rob the trees of valuable carbohydrate reserves for the growth initiation process in July when active growth starts in ber. Improper pruning may also lead to hormonal and nutritional imbalance within the tree and may curtail the productive capacity of the trees considerably.

Training to growers

The ber fruit is borne in the axil of leaves on the young growing shoots of the current year. Hence, a regular annual pruning is necessary to induce a good healthy growth which will provide maximum fruit bearing area on the tree. Pruning is an essential operation to maintain vigour of trees, fruit productivity and yield of ber (Singh *et al.*, 2004). The proper training is given to highlight the objective of pruning, as it is to produce more number of fruits with high quality marketable fruits at a low cost. The training may also include the practice of T-budding, for top-working in wild or inferior trees. Apart from these, pruning also lead to rejuvenation, better ventilation, and higher penetration of sun light and also become feasible in application of plant protection chemicals (Bakhshi *et al.,* 1997). Pruning also lead to rejuvenation, better ventilation, and higher penetration of sunlight, and also become feasible in application of plant-protection chemicals (Bakhshi *et al.,* 1997). Ber growers should have clearcut understanding of pruning and training technique, while going for adoption of rejuvenation technology.

Rejuvenation

Ber trees can be made productive by general pruning and in case of old senile unproductive trees hard core reiterative pruning is required for rejuvenation.

***General pruning*:** Ber is summer dormant and pruning is done during summers. Pruning in ber is prerequisite for better growth and yield of fruits because it bears on current season's growth. Therefore, it is very much essential to ascertain the extent of pruning. Proper pruning keeps the trees productive and improves fruit size, weight and quality. The fruits are borne in the axils of leaves on young shoots of current season. Therefore, a regular annual pruning in ber is very much necessary to induce a good and healthy growth which provide maximum fruit bearing area on trees. Thinning out of branches is also necessary to avoid too much crowding which helps in reducing the incidence of pests and diseases. The best time for pruning is after harvesting during the hot and dry season or when it sheds its leaves and become dormant. Dry, dead and diseased wood, criss-cross branches and weak crotches are removed at the time of pruning. After the establishment of the frame work, pruning should be continued every year. For optimum yield and good quality fruits, 25% of one-year-old shoots should be headed back. Drastic thinning and severe pruning should be avoided, which affects the growth of plant and gives poor yield.

***Hard core pruning* :** Ber trees if not pruned properly every year, start giving diminished yields and small fruits after bearing normal crops for a few year. In unpruned trees, old wood goes on accumulating every year. Such tree become unproductive with weak growth and may get easily infected with diseases like powdery mildew and insect pests like fruit fly, lac insects and termites. The fruit size is reduced and the fruits become mis-shapen, curved, corky and fall prematurely. Such old trees can be revived and rejuvenated by severe pruning in May.

Ber trees need to be rejuvenated after the age of about 25 years (Bal, 2006). In case of old senile unproductive trees, the operation starts with removal of the scaffolds with sharp saw by keeping 10-15 cm (upto 30 cm) long stub/stubs (it can be kept only 4-6 stubs per trees) during May. Select, out-growing 2 sprouts on each stub during July-August the same year. These sprouts grow very fast, hence, the pinching of tops should be done in August-September so that side branch may develop on these sprouts. Thus, full-grown tree is ready up to November- December of the same year. Another approach may be to thin out 12-18 shoots per tree in the fourth week of August. The second thinning of shoots should be done during second fortnight of September (before flowering). Give light pruning during May next year to get some fruits. Such trees can again start yielding with commercial viability from third year onwards and heavy crops in sustainable way for another 30 years.

Top Working

Inferior 'Katha' ber trees growing on neglected lands and along the boarders of the fields yield crops of very small and inferior fruits and do not fetch any income. Such uneconomic and inferior ber trees.(cv. Katha) can be top-worked by budding with superior cultivar. The method of top working in wild ber trees is quite simple and can be practiced by the trained growers themselves. For this, head back the major limbs to keep 15 cm long stubs on one or two scaffolds. Headed back trees produce a number of healthy shoots from below the cut ends during June – July. Of these, two strongest shoots should be retained and remove rest or among many sprouts emerge on the stubs in July, T-budding should be done on 2-3 shoots on each stub during July – August and remove the rest / unwanted sprouts and continue to remove the same for a year. The portion of the shoots above the point of insertion of a bud is cut off entirely soon after the bud begin to sprout. Save the sprouted buds from breakage by giving little pruning initially. Though, the successfully top-worked trees develop sufficient head growth within a year and may bear a small crop of fruits. The newly-developed scaffold system from the sprouted buds come into bearing within two years and then fruit yield and quality enhances in subsequent years.

Manuring

After reiterative pruning, fertilization has to be done during rainy season. Apply whole farm yard manure just after removing the pruning wood in June (Mishra *et al.*, 2011). Urea should be split into two parts. One part should be applied in August and second part during November. Deficiency of N, P and K was found to inhibit shoot length, leaf number and axillary branch production. Regular application of manures and fertilizers is necessary. A well grown tree of 10 years old ber tree 30-50 kg of FYM, 250 g of N (in two split dose ie. August-September, November-December). 250 g of P (in single dose) and 50 g of K/ tree /year under adequate moisture conditions. In western Rajasthan, application of 10 kg of FYM for one year old plantation and then increasing the dose by 10 kg for each year up to 5th year is recommended (Pareek, 1983). Foliar application of 2% urea at pea nut size is found to increase the quality of fruits in terms of improved TSS, and decreased acidity content.

Irrigation

Since it has very deep tap root system and xerophytic nature and is very hardy, once gets established, needs little care and irrigation. It prefers dry period during flowering and needs irrigation for November-February for better fruit

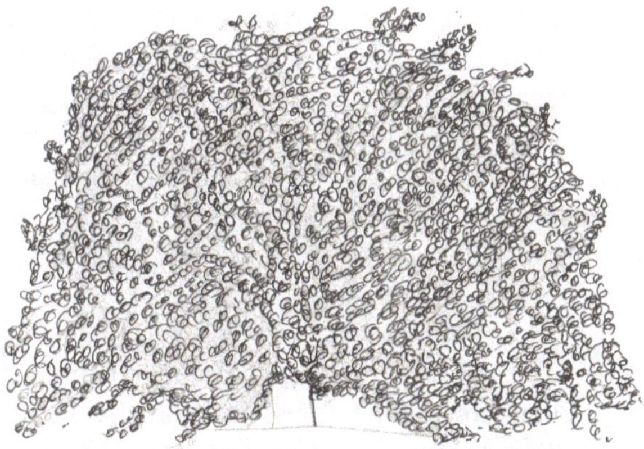

Unproductive old senile tree

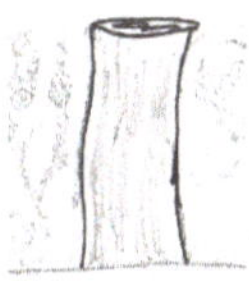

Reiteratively
pruned tree

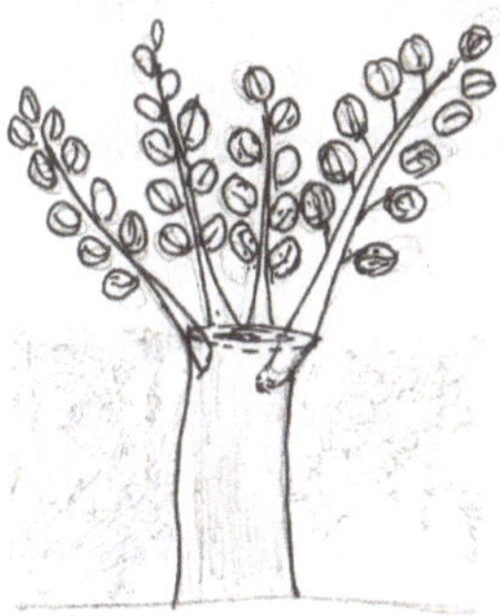

Thinning of shoots

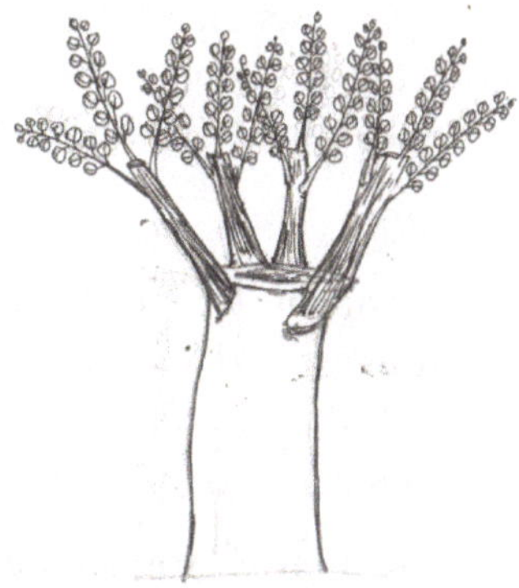

Training of shoots

Rejuvenated tree in fruit bearing

development. No flooding should be carried out till plants cover whole of the space provided. First irrigation should be given in June after pruning for hoeing of the tree basins and for application of farmyard manure. No irrigation should be applied during rainy season. If there is drought, apply irrigation needed after fruit setting. Thereafter, give irrigation only at monthly interval up to February. Fruit start ripening from March onward and harvesting is almost complete in April in Punjab. No irrigation should be given during and after fruit harvest.

Intercropping

The vacant land created after hard core pruning with stub remaining, should be judiciously used by growing compatible crops. No crop should be grown in *kharif* season, since it will need irrigation during dormancy of ber plants. Grow 'rabi' crops like wheat, gram, mustard, sarson, toria, barseem, etc. For the first two years, later on grow only toria or winter vegetables so that the field is free during March-April for easy harvesting of the fruits. Exhaustive and tall growing crops like bajra or jowar should not be grown at all.

Mulching

Cover cropping with Kulthi (Dolichos biflorus) was found to increase WHC of light soils as a result of increased OC content in arid region in the tropics (Pareek and Vishal Nath, 1996). Mulching under the canopy area prior to fruit development stage is always beneficial in conservation of moisture.

Management of Physiological disorder

Fruit drop

Fruit drop may be due to any one or combinations of many causes like disintegration of ovules, Shrivelling of fruits and High incidence of powdery mildew. Fruit drop depends on cultivar, maturity period of cultivar (early, mid or late) and extent of fruit set, and bearing of tree. Fruit drop has been found to be reduced by : i, the application of either 2,4-D (10-20 ppm) or 2,4-5-T (5-10 ppm) or GA3 (25 ppm) in cultivars like Kaithali and Umran; ii, irrigation of ber trees after fruit setting; iii, apply 50 kg FYM, 250 g N + 50 g P + 50 g K/tree after pruning and fruit setting; iv, the application of NAA @ 10ppm at fruit bloom is found to improve the fruit size, while the foliar application of Borax and zinc sulphate is found to improve quality of fruits.

Management of insect and pests

Fruit fly (*Crpomyia vesuviana* Costa)

It is a specific pest of ber both on cultivated and wild varieties and is active during January to March under north Indian conditions. The symptoms appear on fruits as gullies inside the fruits, deformed brown coloured, rotten fruits. The attacked fruit assumes rugged appearance and does not readily fall on the ground.

Removal and destruction of infested plant parts, and follow of clean cultivation/ sanitation of orchards. Ploughing soil May - June and August. Spray Rogar (0.05%) in February – March. Grow resistant cultivars, Illachi, Umaran

Bark eating caterpillar (*Indarbela quadrinotata*)

Galleris of frassy web near the fork of tree branches. Painting the holes with 0.1% monocrotophos is advised. Use small pieces of cotton, soaked in solution prepared of 100g of soap + one litre of Kerosene oil, for plugging inside the hole and seal with mud. Resistant cultivars - Desi Alwar is good to control caterpillars.

Hairy caterpillar (*Euproctis fraternal*)

Feed on leaves, fruits and tender shoots. Dusting of 10% Lindane (or) 0.1% Carbaryl (or) Endosulfan

Lac insect (*Kerria lacca*)

Ber is one of the host plant of this pest. Suck the sap of branches and scale like appearance (lac) of twigs and branches. The main season of infestation - April - May and September in states like Bihar, Madhya Pradesh and Punjab. The fruit yield is drastically reduced.

Remove and destroy infected twigs / branch (dry parts). Spraying of Rogor (Dimethoate) @ 0.25% or Dimecron, after scraping off of the infested twigs, helps control.

Fruit borer (*Meridarches scyrodus*)

Feed on fruit – pulp, leading to fruit rot. Apply Monocrotophos - 0.2% twice.

Leaf defoliating beetles

They feed on leaves during night. Season of infestation is June - July more in neglected orchards. Spray Thiodan 0.05% (or) Monocrotophos and plough the soil and expose the grafts.

Management of Diseases

Powdery mildew (*Microsphaera alphitoides* f. sp. *ziziphi*/*odium* spp.)

It appears at the time of fruit setting in all the commercial cultivars causing heavy losses. The symptoms appear as leaves shrinkage and defoliation, fruits become dark brown, corky, canker, misshapen, under developed. The fungus survives on wild ber and certain rootstock species and spraying of Karathane (0.2%) or Bavistin (0.05%) or sulphur dust @ 250 g/ tree, keeps it under control. Resistant cultivar, Chhuhara, Gola and Thornless should be grown.

Sooty mould (*Isariopsis indica* var. *zyziphi*)

This disease lowers leaves of plant, resulting severe defoliation. Symptoms as black spots on lower surface of leaf, drying in Sept-Oct. Under humid weather conditions fruits may also get affected.

Spray Dithane (Z-78 0.3%) or copper oxychloride (50 WP-0.3%) at fortnightly intervals November onwards.

Leaf spot (*Alternaria* sp.)

Necrotic spots develop on leaves and drying in January-February. Use Dithane Z-78 0.3% or Copper-oxy-chloride (50 WP-0.3%) at fortnightly intervals starting from November.

Rust (*Pakospora zizyhivulgaris*)

The infestation starts as irregular (brown to black) rust on lower surface of leaves leading to pre-mature defoliation. Spraying of Bavistin (0.1%), 2-3 times. Growing of resistant cultivars, Gola and Banarasi is advised.

Fruit rot (*Phomopsis*, *Colletrotrichum*, *Alternaria* spp.)

The symptoms occur as brown spot on distal end of fruit. Spraying of Dithane (Z-78 0.3%) or Blitox (0.2%) 2-3 times at 7-8 during initial stage of fruit development is very effective.

Fruit yield

Quality fruit yield in the tune of 80 to 100 kg per tree starts from third to fourth year onwards in rejuvenated old trees.

Shoot pruning primary level

Shoot pruning at secondary level

Hard core pruning for rejuvenation

Chapter 10

Cashewnut

Cashew (*Anacardium occidentale*), is an evergreen tree in the family Anacardiaceae. It is grown for its edible fruits (nuts). Cashew originates from North-East Brazil (Salam and Peter, 2010). As a fast growing woody perennial, is characterized by spreading branches. Cashew trees can reach a height of 12 m and have an economic lifespan of 25 years. A Canopy of cashew develops irregular if not checked properly. In later stage, its canopy shape and size become difficult to manage (Nalini *et al.*, 1999). Manipulation of canopy size and shape by means of pruning and training for its containment and better yield was not felt as a necessary orchard management technique for cashew. Majority of the older plantations are of seedling origin of non-descript material and have poor genetic potentiality. They have become senile. Such plantations can be rejuvenated provided they are managed properly and not very old (Gurupasand *et al.*, 2012). If the plantations are of improved varieties having good genetic make-up such plantations can be rejuvenated by limb pruning or if they are genetically low yielders they need to be top-worked by beheading and grafting with better material for improving the yield performance.

PRESENT STATUS

India is the second largest producer and exporter of cashew at the global level. Kerala accounts for 15 % of cashew production in the country. Among Indian states, Andhra Pradesh stands first in cashew area (0.171 million ha), whereas Maharashtra stands first with respect to cashew nut production (0.210 million tonnes) during 2007-08 (Salam and Peter, 2010). Cashew processing units of Kerala, which share more than 90% of the total capacity at the All India level, finds it extremely difficult even to maintain one-third of its capacity. Processed cashew kernels have good demand in international markets. Promotion of high density planting with high yielding cashew grafts is imminent. In India, Goa has more land (unit area basis) under cashew than any other state (Salam and

Peter, 2010). The R&D activities for cashew is on with fast pace and the area under this crop is also increasing.

PRODUCTION CONSTRAINTS

Most of the old existing orchards are of seedling origin. The existing cashew trees gives small nuts, which did not meet the market demand. In general, except some weeding for orchard floor cleaning, no other cultural practice is practiced in most of the cashew orchards because very few of the orchards were intercropped mainly with some short duration annual crops. With the advancement of age, cashew trees became old and senile due to negligence, and therefore succumbed to pests and diseases too. As senility developed, production dropped tremendously or stopped completely. Because of this, farmers abandoned the cashew orchards rendering the trees look like wild trees. Hence, productivity of cashew suffers a set back in entire country due to predominance of old senile and unproductive trees. The growers, then prefer to replace it with new plantation. In commercial bearing orchards, canopy of cashew has irregular canopy shape and size which are difficult to manage and thus result in poor nut yield in the later years as the lower branches of canopy gradually die and become empty. Such plantations need replanting with high yielding varieties.

Cashew often grown on marginal soils and also on wasteland is mostly unsuitable for other economic crops but responds nursing. The tree requires good drainage, friable soils and can tolerate a wide pH range and even salt injury. It grows usually unirrigated but responds to summer irrigation (during initiation of vegetative growth of current season). Cashew do not withstand water logging and proper drainage is essential. Systemic plant protection schedule is not being followed by most of the cashew growers, as to get steady returns from the cashew plantations. Although, magnitude of the physiological disorders is rather comparatively low in cashewnut, however, yellow leaf spot and little leaf are common problems and cause yield decline. There is a need for establishment of factors behind the decline in cashew nut production in our country.

Need for rejuvenation

Due to lack care and management, by mere ignorance of the growers. The productivity of cashew has been declined drastically in the existing plantation in entire country. Since most of these trees are of seedling origin, have become old, senile and unproductive. Due to ignorance, growers, then prefer to replace it with new ones. Which can devoid them from the benefits of rejuvenation technology (including top-working). As top-working using improved varieties and other improved knowledge and technologies, many unproductive/abandoned cashew farms have been rehabilitated and/or upgraded in productive ones. As top-working is a better solution to raise the level of productivity of cashew trees. Top-working is nothing but replacing the crown of the tree with the shoots of high yielding varieties. Hence, rejuvenation of senile plantation should be done at priority basis.

REJUVENATION TECHNIQUE

Identification of old orchards

Cashew trees in orchard, if left unattended, grow wild, haphazardly with more horizontal and downward bending way and as the age advances lower portion becomes devoid of leaves, shoots and branches. These are attacked by many pests, diseases and disorders. These type of orchards become poor and irregular bearer and ultimately become unproductive. These orchards should be identified for reiterative pruning and followed by top working for rejuvenation.

Training to growers

Farmers still treat cashew trees as forest trees. They rarely apply manures and fertilizers as well as hardly undertake plant protection measures to maintain the productive nature of the tree. They are still ignorant about the invasive nature of rejuvenation technology including the cumbersome process of top-working operation. The growers need proper training for maintaining the health and vigour of the tree for potential production. There should be onsite theory and practical training (including demonstration trial) for convincing the grower to develop courage, confidence and skill to go for adoption of the technology. Knowledge sharing and upgrading are very vital in any programme formulation, planning and successful implementation. The training or interaction meeting with growers groups/farmers for exchanging ideas with experts on the declining factors mainly about devastating diseases is must to facilitate the transfer of technologies. The benefits of existing root system and precocity due to adoption of rejuvenation technology should be made understandable to them for faster adoption level. Various activities took place to inform and sensitize the actual beneficiaries for developing the convincing level and implementation plan, as the activities included meetings, seminars and training sessions. Various cashew growers received training on cultural practices and were introduced to various cashew knowledge, technologies and skills such as grafting, nursery management, apart from upgrading of old cashew orchards and formative pruning at various locations in phase manner.

Pruning for rejuvenation

Limb pruning technique can be used as a good tool to rejuvenate unproductive plantations. The trees of more than 15 years, which cover entire inter- and intra-space and show decline in yield or fail to give yield (Rajanna and Nayak, 2007). This technique involves cutting of the primary branches from the main trunk during August-September so that, new sprouts commence in the next month which subsequently flower and fruit in January-February. Pruning should be done by using a saw to achieve a perfect cut and end without any cracks or damage to the bark of the branch. This technique has given successful results for cashew trees cultivated in fertile soils in plains of Karnataka.

The normal optimum spacing recommended for cashew is 8m X 8m. There is overlapping of canopy in 8 to 9 years of age depending on the nature of variety unless the canopy is maintained by trimming and thinning. Plantations which have overlapping branches will lead to dead wood development and decline in yield in later years. Such plantations can be rejuvenated through canopy redevelopment by limb pruning.

Rebuilding of canopy is important in older plantations, where the canopies are over crowded, resulting in reduction in yield (Nayak, 2011). This is possible by heading back (limb pruning) the exhausted canopy at 1.0-1.5 m height. Immediately after beheading the trees, the pruned wood needs to be removed from the plot so as to prevent the damage by shoot hole borers. Remaining portion of the trunk and exposed part of roots is to be swabbed with 1 per cent Chlorpyriphos at least 2-3 times after beheading till new canopy is developed. Regular checking for the damage by cashew stem and root borer is to be done and managed. This cut ends of the branches should be smeared with 10 per cent Bordeaux paste to prevent the gummosis and entry of any pathogens. The best season for limb pruning is May-June i.e. immediately after the harvesting of the annual crop.

The new flush arise from the dormant buds on the trunk. These new sprouts from the canopy within a period of 6-7 months depending on the variety. Yield level during the first season after pruning are slightly lower while the yield from second season onwards is better than the un-pruned plants. Yield performance of the 4 varieties during first 2 years after limb pruning has given satisfactory results.

Good phyto-sanitation is to be adopted to manage the rejuvenated plants. The newly developed canopy needs to be protected from Tea Mosquito Bug (*Helopeltis antonii*) and leaf beetles as they are also more vulnerable for the damage in the initial 5-6 months of canopy development. Care should also be taken to treat the cut wounds with 10%. Bordeaux paste during the beginning of monsoon season to prevent from any fungal infection and gummosis. If the plants are properly protected the canopy redevelopment can help in enhancing the nut yield to the fullest potential after 2-3 years. The limb pruned trees be pruned or trimmed as and when required in subsequent years. It may be noted that the yields level of limb pruned trees are much better than the freshly planted cashew grafts of comparable years due to better root system. Pruning operation should be taken up only after complete picking of nuts which coincides with temporary quiescent stage in cashew (May-June in West Coast).

Rejuvenation by top-working

Rejuvenation of old and unthrifty cashew trees by top-working, by grafting with scions of better variety has been attempted in early eighties at Agricultural Research Station, Ullal of the University of Agricultural Sciences, Bangalore (Rao, 1985; Guruprasad *et al.*, 1988; Khan *et al.*, 1988) and subsequently modified as per local requirement and popularized by National Research Centre on Cashew, Puttur and several other cashew research centres in the country.

Though the top-working is a promising technology to rejuvenate the old senile plantation of seedling progeny in a short period, the technology could not be adopted on a large scale due to problem of cashew stem and root borer (CSRB), which damages and kills the top worked trees. The technology is being recommended only for the homestead gardens of young age which can be properly monitored and maintained.

The points to be considered are :

Age of the tree : Tree should be within 15-20 years age. As far as possible it should have smooth bark and should not have many crevices on the bark at collar region. The colour of the trunk should be brown or ash coloured and not black coloured.

Protection from CSRB attack : Tree should not have been damaged by CSRB, as care should be taken that either the trunk portion or exposed root portion on the ground should be free from stem borer damage. The symptoms of CSRB can be ascertained from the oozing of gum or the powdery mass (frass) fallen near the entry point of CSRB.

Tree health : Trees should be healthy enough with spreading lower branches. It should have well developed root system and should have a capacity to cope up growth of new canopy. It should be free from other pest and diseases.

Top-working

Top working involves two steps as follows :

Beheading of the tree : The most suitable time for beheading of the tree is April-May i.e. immediately after the harvest of the crop in West Coast region.

Old senile unproductive cashew tree

Pruned tree

Emergence of shoots

Shoot thinning

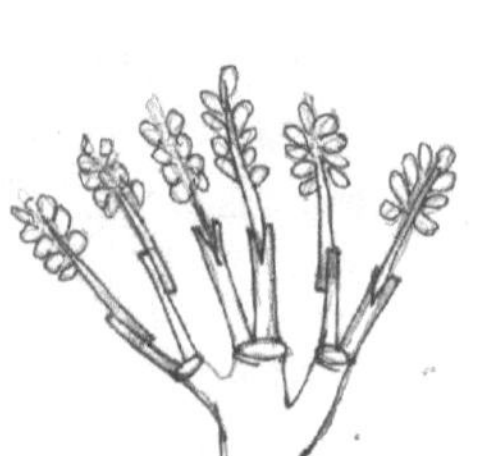

Top grafting

Rejuvenated tree in fruit bearing

It also depends on the fruit bearing pattern of the tree (early, mid and late season) in different varieties. After selecting the tree for beheading, cut the limbs and reduce the weight of the crown and finally retain the stumps of the plant at 1.0 – 1.5m height. While pruning the limbs, care should be taken so that the bark should not have split or damaged on the retained stump. Immediately after the beheading of the tree treat the cut wounds with 10 per cent Bordeaux paste so that they should not be any gummosis and fungal infection in the wound. After beheading tree the cut plant material should be cleared from site immediately to prevent the harbouring of pathogens and insect pests.

Grafting : Grafting new flushes emerging from the trunk of the cut tree can be taken up after 45 to 60 days onwards. About 200 to 300 shoots emerge from the trunk below the cut end depending on the size of the remaining trunk. On a cut tree about 15-20 shoots of different branches in different directions may be selected and grafting by soft wood grafting technique may be taken with the scion material of required or elite varieties. The period between June and August is most suitable for top working under west coast region. After successful grafting the shoots remaining un-grafted, need to be gradually removed in phased manner. After about 4-5 months of grafting only about 10-12 successful grafts in the different branches in different directions may be retained and rest may be removed. The new flushes emerging from the trunk subsequently also be frequently removed. Initially grafts or shoots may be protected from damage by wind and animals by providing staking and support with sticks. From second year of graft, the plant will grow into a tree and give flowers and fruits like normal young tree.

In plantations of young trees of 8 years age of variety Vengurla-2 when top worked with the scion of Venggurla 4 to replace the variety in scion bank, 88 trees out of 94 top-worked trees could be successfully saved and managed along with economics. The large scale plot trials on top-working conducted both at NRC on Cashew and Karnataka Cashew Development Corporations, revealed that utmost care is required to protect the top-worked tree from the devastating pest of cashew stem and root borer (Salam and Peter, 2010).

Nutrition

A fertilizer dose is recommended for a commercial orchard tree i.e. a dose of 750 g N, 325 g P_2O_5 and 750 g K_2O/tree (Janik and Paull 2008, Veeraraghavan *et al.*, 1985) should be used in rejuvenated orchards too, but care should be taken to apply the fertilizers after a month gap of reiterative pruning operation is over, and that too with half of the recommended dose first year and the remaining second half in second year same time, through ring method of application. The normal recommended dose should be followed from the third year onward as yearly basis (Kumar *et al.*, 1998; Lefebvre, 1973).

Irrigation

Cashew is mostly cultivated as a rainfed crop. However, provision of protective irrigation for proper vegetative and reproductive growth responds positively. After a gap of 25-30 days in case of reiteratively pruned trees, irrigation is very much required. During the third year of rejuvenation, irrigating 80 litres of water / tree once four days through dippers as adult tree dose from December to March can also bring bearing and increase yield (Lenka and Pattnaik, 2008).

Mulching

The open space created after hard core pruning increases the surface evaporation and reduces the moisture content in the rhizosphere of the pruned trees of rainfed cashew. Mulching is an important soil-moisture management strategy, for sustained moisture conservation practices and also control of weeds. Conservation of moisture during hot dry months before rainy season and from December to March brings fast vegetative growth, starts early bearing and quality fruit yield in cashew.

Intercropping

Intercropping in rejuvenated orchards with suitable leguminous crops such as black gram, green gram, hoarse gram, pigeon pea etc., can be found profitable and beneficial in rejuvenated cashew orchards. Other intercrops of preference are groundnut and vegetables (solanaceous and cruciferous groups). Pineapple is the most profitable intercrops in the cashew plantation. Hence, the open space created after reiterative pruning, should be utilized carefully with profitable intercrops. Ginger, lemongrass and tapioca are also found suitable intercrops in cashew. Any intercultural operation is initiated after 25-30 day of pruning operation is over (Lenka and Pattnaik, 2008; Janik and Paull, 2008).

Management of Physiological Disorders

Little leaf

Little leaf of cashew is caused due to the deficiency of Zinc. The leaves remain small in size giving rosette appearance. Application of zinc sulphate (0.5%) to the soil or by foliar spray means is helpful in preventing the disorder.

Yellow leaf spot

The occurrence of yellow leaf spot, an enigmatic disorder in cashew, is found to be associated with low soil pH (4.5-5.0). The appearance of yellow leaf spot might be due to molybdenum deficiency. The affected leaves are deficient in molybdenum and had excessive amounts of manganese.

The trees sprayed with 0.03 per cent solution of ammonium molybdate once during pre-monsoon (June) and another after end of monsoon (September) is found to be effective in receiving the plants. The disorder can also be controlled by correcting soil pH with lime.

Management of Pests

There should be adequate attention to conduct research on diseases and insect pests that addresses the needs and aspirations of the farmers in order to solve their cashew production problems in the field. As even after reiterative pruning and/or followed by top grafting for rejuvenation, there is very high mortality in the field and the purpose of rejuvenation becomes failure.

Tea mosquito bug (TMB) (*He/opeltis antonii* S.)

Incidence of Tea mosquito bug on cashew occurs at new flushing, flowering and *fruiting* stages during September to April *and* causes crop *losses ranging* from 20-60 per cent. The damage is caused by adult bugs as well as by immature stages (nymphs) which suck the plant sap from the tender flushes and succulent shoots, flower panicles, growing nuts and apples.

Dark brown patches are seen tender stems and inflorescence rachis. Severe incidence eads to flower blight and drying of young shoots and flower panicles. Attack on tender nuts and apples causes shrivelling of developing nuts leading to immature fruit drop. The affected trees will conspicuously lave the dried leaves on dead tender shoots md dried inflorescences on tree canopy which can be prominently seen from a listance. Severely affected branches may lead to the secondary infection by fungus *(Botrydiplodia theobromae)* causing 'die lack' disease, i.e drying of branches. Such branches will have the brown discolouration in the sap wood extending towards lower portion.

This pest can be effectively managed by spraying the trees in three critical stages. 1, Establish the cashew plantation by planting 2-3 high yielding varieties with varied degrees of tolerance or escape mechanism so that the impact of pest incidence will be reduced due to varied degree of pest host reaction; 2, spray the tree canopy at flushing stage with Monocrotophos at the rate of 1.5 ml/l of water (0.05%) or with *Lambda-cyhalothrin* at the rate of (0.003%); 3, spray Carbaryl (50% WP) at the rate of 2g/l of water (0.1%) or Chlorpyriphos (0.05%) at flowering stage; 4, repeat the 'first spray' at initial fruiting stage; 5, Cut the 'die back' affected branches up to the healthy region and apply the cut end with Bordeaux paste (10%) and spray the tree with 1 per cent Bordeaux mixture solution and 6, arrest the build- up of the pest population by maintaining the hygienity in plantation and also care should be taken for controlling the population even on the alternate host trees.

Cashew stem and root borer (CSRB) (*Plocaederus ferruginneus* L.)

This is the most dreaded pest of cashew trees, the immature stages (grubs) of which feed on the bark of trunk and roots and out rightly kill the tree within few days. Due to its concealed nature of damage, it is difficult to make out the status or level of damage to the tree and therefore, unlike TMB, is a threat to the life of the tree. This is a dark brown beetle with long antennae which lays eggs in the crevices of the bark on the collar region.

Incidence of the pest can be made out by the presence of extruded frass (like coarse saw dust powder) on the ground surface at the base of cashew tree trunk. On closer examination, one can see small holes on the collar region through which the frass is pushed out by the insect. After hatching of eggs, the grubs immediately start boring into bark of collar portion and root for feeding on the inner portion of the bark tissues. Further, they bore irregular tunnels throughout. As a result of severe injury to the vascular bundles, gum like substance oozes out and gets deposited on hardening to form dark brown clumps on the affected region. In case of severe infestation, due to continuous feeding on vascular tissues, the trunk becomes internally girdled. At this stage, leaves show yellowing symptoms, start shedding immaturely and subsequently the entire tree dries up within a few days. Affected trees also tilt on one side due to loss of anchorage, if the injury is severe on anchoring roots. Infestation occurs round the year.

The integrated management of this pest require 1, regular monitoring for the presence of the frass near the base of the trunk; 2, mechanically removing of the grubs from infected trees in the initial stage of infestation; 3, inject 0.36 per cent Monocrotophos by the padding method; 4, soak/drench the bark of the infested trees with Chlorpyriphos (0.1%), up to one meter above the ground and also by drenching the soil/root zone in a radius of 1.0m around the main trunk to kill the grubs/pupae in the root zone; 5, repeat the injection of all infested trees with Monocrotophos after 15-20 days; 6, it is very important to adopt the phyto-sanitory measures like removal of dead trees infested with CSRB and bury or burn the same to avoid further spread of infestation.

Leaf and blossom webber (*Lamida moncusalis* L.)

As indicated by its name, this pest attacks the young leaves of new flush and inflorescences. The caterpillars form the web of young leaves and flower panicles and feed on these from inside as a result of which, the webbed shoots totally dry up.

Leaf miner (*Acrocercops syngramma* M.)

Larvae of this insect mine through the young leaves of new flushes. In the mined laminal areas the epidermal layer gets separated from the leaf tissues in patches giving white blistered patches.

Apple and nut borer (*Thylocoptila* sp.)

The larvae of this pest bore the region of juncture where the nut is attached to apple and start feeding on tissues. The incidence can be conspicuously seen by the presence of frass and excreta of the larvae at the tunnelled place.

Mealy bug (*Ferrisia virgata)*

Nymphs and adults of mealy bugs suck large amount of sap from the growing shoots, inflorescence and developing fruits. These insects also excrete sugary substance (honeydew) on which sooty mould develops which impairs photosynthetic activity rendering 16-30% losses in nut yield under severe out break conditions. This minor pests can be taken care of by the prophylactic sprays adopted for the management of tea mosquito bug.

Management of Diseases

Powdery mildew (*Oidium anacardii* Noack.)

Powdery mildew disease of cashew is characterized by the formation of an ash-white covering on the surface of attacked tissue. The pathogen infects young growing tissue on all aerial parts of the cashew tree, including shoots, leaves, flowers, apples and nuts.

Powdery was basically controlled by use of sulphur dust, but due to likely environmental problems, alternative fungicides should be tested for effective control.

Leaf and nut blight disease (*Cryptosporiopsis* sp.)

Leaf and nut blight disease is a fungal disease caused by The disease symptoms portrays angular lesions, dark tan with a dark reddish brown margin, often vein limited and containing conidiomata. Lesions subsequently enlarge and coalesce causing blighting and defoliation. Older lesions become papery, silver/grey in colour and develop shot-holes. During fruit setting, infection on young nuts causes rapid blackening and abscission, resulting in significant yield losses. Infection of older nuts results in a characteristic dark, slightly sunken, 'tar spot' like lesion that frequently extends onto the apples. The disease is best controlled by a combination of approaches including cultural, resistance and use of fungicides.

Anthracnose (*Colletroctichum gloeosporoides)*

Anthracnose a fungal disease, which attacks all young and tender vegetative organs together with nuts and pseudo-fruits/apples. Infected shoots as well as flowers dry out cling-on and appear shriveled and curled at the edges. The disease is best controlled by a combination of approaches including cultural, resistance and use of fungicides.

Dieback disease (*Phomopsis anacardii)*

Dieback disease is a fungal disease. However fungus infestation may start following the damage caused by insect attack (*Helopeltis* spp *or Pseudotheraptus wayi*), which are considered as predisposing factors to this dieback infection. The fungus attack young and tender shoots and flowers. The disease is also best controlled by a combination of approaches including cultural, resistance and use of fungicides.

Cashew wilt (*Laetiporus* and *Ganoderma* spp.)

There are two types of macro fungi (*Laetiporus* and *Ganoderma spp)* are commonly associated with cashew wilt. Currently, farmers are advised to remove all *Laetiporus* fruiting bodies as soon as they are seen around the vicinity of target trees. It is advisable that farmers should routinely inspect their trees with effect from onset of rains, so as to remove the fruiting bodies in good time. It has also been found that drought was the cause of cashew wilt syndrome disorder.

Fruit Yield

Harvesting of cashew in Goa is done generally from March to May. Fully matured apples are allowed to drop down naturally with the attached nut. Then the nuts are separated from apples. Nuts are sundried for 1-2 days and stored before selling off in the market.

Old unproductive cashew tree

Declining tree

Declining tree Dried / dead tree

Dense tree canopy

Dense branching

Unproductive shoots

Developed canopy

Profuse flowering

Fruiting in rejuvenated tree

Chapter 11

Citrus

Genus Citrus (*Citrus* spp.) belongs to family Rutaceae and sub-family Aurantioedae. Citrus fruits are cultivated over a wide range of agroclimatic conditions because of their ability to adapt to different situations. Citrus is considered to be one of the most remunerative fruit crops in India, having a lasting niche in the international trade and world finance. Most of the Citrus species are believed to be native to tropical and subtropical regions of southeast Asia, particularly, northeastern India and the region between China and India (Ghosh, 2007). Trees belonging to genus Citrus are comparatively small in size (other perennial fruit crops) and evergreen in nature that are grown in tropical, subtropical and even mild temperate climates. Citrus fruits are non-climacteric so they do not ripen further once removed from the trees. So it is important that they are picked at the right stage of maturity.

PRESENT STATUS

Globally, India ranks fourth in terms of production but in terms of productivity, it figures very low. The current average productivity of 10.9 tonnes/ha compares no way with the average productivity of Citrus orchards of even some of the southeast Asian countries like Japan where high density planting on dwarfing rootstock alone has contributed immensely, revolutionising commercial cultivation of Citrus crops. With growing economy, awareness towards quality, nutritional safeguard and consequently higher demand for quality fruits and beverages, the challenges are increasing for enhanced quality and productivity (Kahlon, 2008). These emerging challenges call for paradigm shift in production with up-to-date state of the art technological interventions (NRCC Vision 2030). Citrus comprises many species of economic importance. In India, a few species are commercially cultivated, which include grape fruits, lemons, limes, sweet oranges, and mandarins (Singh and Singh, 2003).

PRODUCTION CONSTRAINTS

Citrus is grown under varied climatic and geographical zones of India. Its cultivation faces a variety of problems leading to 'citrus decline' which has been rampant in major citrus growing regions (Srivastava and Singh, 2008). The reasons may be many such as unhealthy planting material, unsuitable soil, malnutrition, lack of irrigation facilities, insect pests, disease etc. In commercial bearing plantation, lack of proper cultural practices restricts exports of its quality fruit production. Citrus fruits further suffer from a number of different physiological disorders that may considerably affect the fruit crop by destroying the trees and /or the fruits. The major problems include fruit splitting, fruit drop, granulation, tree decline etc. Non-availability of proper technical knowledge of disorders is the basic reason for low production, short productive life and gradual decline of Citrus trees. Citrus production is severely hampered due to disorders across the world. Citrus decline can be termed as a syndrome. There is not a single cause of its occurrence, but is the effect of many causes, which may vary from field to field or area to area. Low productivity or un-productivity in Citrus are also caused by the syndrome. Some of the factors causing complexities are due to soil characters, drainage, salt concentrations, water logging, frost injuries, etc. Over and inadequate fertilization is one of the reasons of citrus decline (Srivastava and Singh, 2008). In India, Citrus fruit crops flooded by smallholders and even commercial farmers without any segments. This type of irrigation may save some labour, but is very injurious to root system, ultimately affecting plant growth and fruit yield. The root rot is caused by over irrigation. However, its production is seriously constrained by various diseases including *Pseudocercospora* leaf and fruit spot. The vigorous vegetative growth resulting in crowding and shading reduces yield and fruit quality, and makes difficult grove access, maintenance of irrigation, fertilizer delivery, spraying and harvesting.

Need of rejuvenation

The Citrus decline has plagued our plantations invariably in space and time warranting for massive rejuvenation immediately. Citrus growers are still not aware of the available technology options for raising the productivity from present level on sustained basis. However, adopting an appropriate and effective canopy management strategy (CMS) solves these problems, and prevents trees investing energy in misdirected or poor quality growth (Krajewski, 1996; Krajewski and Pittaway, 2000). Fruit quality originates in the groves. It is strongly affected by cultural practices, and pruning (Krajewski, 1996; Krajewski and Pittaway, 2000). Since light becomes a limiting factor in crowded groves, pruning improves light accessibility. Adjustments must be made in tree height, row middle width and hedging angle to maximize sunlight impingement on the tree canopy. Sunlight not only influences flowering and fruit setting but also enhances fruit quality and colour development. The foliage in Citrus trees therefore acts as an important food storage area, and heavy pruning causes the tree to produce vegetative growth at the expense of fruit production. Vigorous shoots called water sprouts or suckers show extreme apical dominance with no side-branch development. Apical dominance varies somewhat with vigour and variety, which can only be controlled through pruning. Removal of terminal bud destroys apical dominance so that one or several lateral buds commence to grow and branching results can be regulated for desired outcome.

Techniques of rejuvenation

Rejuvenation in Citrus declining orchards can be done by adopting the practice of pruning in different intensities depending on the age of the tree and severity of decline. A pruning programme should begin before any heavy cutting is necessary and should be continued at appropriate intervals so that desired tree (hedgerow) size and shape can be maintained at low cost and with minimum loss of canopy. A regular pruning programme eliminates the necessity for a large cuts, avoids excessive vegetative growth, maintains good fruit production. One should be careful about the time of pruning operation to be followed. The best time of the year for this practice depends upon variety, location, severity of pruning and availability of equipments, but any how it should not be practiced during rainy season.

Identification of old senile orchards

Many citrus growers have adopted closer planting to achieve higher fruit production. But non-adoption of proper package of cultural practices have led the orchards dense and in declining stage. These orchards need to be identified for corrective pruning and training operations. The trees of orchard, which have showing declining phase due to loss of vigour results in falling of the foliage, low yield with smaller fruits, start dieback of twigs. Such orchards need rejuvenation on time (Singh *et al.*, 2004).

Training to orchard growers

The growers are primarily be made acquainted with the good management practices and its timely adoption. The pruning operation is highly skilled operation and the response of pruning depends on variety, tree age, tree vigour, fruiting habits, growing conditions and production practices. As no one system or set of rules is adequate for the numerous situations/conditions encountered by the growers in the field. The pruning of fruit-bearing trees should commence when the trees start touching each other. The hedge row system of pruning should be advised to ensure regular cropping every year by pruning alternate sides every year with prophylactic measures, so that infestation of tree decline should not take place. In this regard, proper onsite theory and practical oriented training to the growers are required, which encourage them to acquire proper understanding of the principles involved in pruning and to take advantage of research results and knowledge to develop skill through custom operators observations.

Pruning

The pruning of citrus trees results in production of vegetative re-growth, which must be carefully protected from all diseases and pests. Pruning healthy, mature citrus trees usually reduces yield in proportion to the amount of foliage removed and can delay fruiting in young, and non-bearing trees. Pruning should therefore be limited to that required for future canopy bearing surface development and for the conduct of efficient cultural and harvesting operations, and to save from the declining phase (Oostrum, 1962; Wheaton *et al.*, 1978; Reddy and Patil, 2004; Srivastava and Singh, 2008). Heading back and thinning out are the main types of pruning cuts and are used for somewhat different objectives. Heading back removes the terminal portion of a shoot or branch, destroying apical dominance and stimulating lateral bud breaks. This tends to produce a more bushy, compact tree. Mechanical hedging and topping are forms of mass heading back. As the individual tree or hedge row increase in size, internal wood may become less productive and eventually die. Thinning out involves the removal of complete branches to laterals or to the main trunk and is done by selective pruning with hand-held equipment. It encourages longer growth of the remaining terminals and can result in a larger and more open tree. This type of pruning may be done for better light penetration in the tree, but is generally considered labour intensive and not much practiced in Florida.

Young Trees : Severe pruning and training of young, non-bearing trees tends to delay in fruit production and should be avoided. Most trees usually need no pruning for the first few years in the grove except for removal of sprouts on the trunk. These can be easily brushed off when trees are young and tender and the wounds are small. Larger sprouts should be cut off flush with the trunk. Sprouting on the trunks of young, non-bearing trees can be greatly reduced by using a commercial sprout inhibitor containing NAA. Suckers arising from rootstocks should be removed early. When the tree is 3 or 4 years old, depending on its growth, branches that are too closely spaced or are crossed and entangled may be removed. This pruning should be light, just sufficient to establish a desirable framework without stimulating excessive vegetative growth. Optional pruning during the next several years should be limited to removal of water sprouts, dead wood and occasional branches which interfere with the growth of scaffold limbs.

Mature Trees : The mature trees require little pruning until they approach containment size. Deadwood may be removed every 3-5 years, depending on the amount present and on the labour supply. Growers need only remove dead branches of one-fourth inch in diameter or larger since smaller dead wood is often broken off in harvesting and other operations. All cuts should be made into live wood, flush with a larger branch or at a lateral. When crowding occurs in case of HDP, resulting in inadequate light conditions, loss of foliage and fruit production takes place in the lower portion of the tree. Cultural and harvesting operations are also adversely affected (Gonzalez *et al.*, 1996; Lewis *et al.*, 1963).

Hedging : Hedging, which consists of cutting back the sides of trees to prevent or alleviate crowding, is a common practice in important citrus-growing countries (rarely in India), since the development of mechanical equipment for this purpose. Hedging leaves numerous cut wood surfaces along the side of the tree canopy from which new sprouts arise eventually developing into a wall of new foliage (Cary, 1981; Castle, 1978; Wheaton *et al.*, 1984; Reddy and Patil, 2004). The closer the spacing and the more vigorous the trees, the sooner hedging is required and the more frequently it needs to be done. Removal of a large portion of the tree may be required when pruning is deferred until severe crowding occurs. However, great majority of new plantings which are designed to provide a continuous hedgerow as rapidly as possible, are apparently being successfully managed and harvested.

Topping : Topping should be done before trees have become excessively tall. It should be an integral part of the maintenance programme. Long intervals between topping increases the costs of the operation due to heavy cutting and thinning. Excessively tall trees are more difficult and expensive to harvest and spray. Topping (in reiterative manner) trees increase light penetration into trees, thereby stimulating intense growth and flush (Khan and Shyamal, 2004). It also reduces harvesting costs, enhances pest and disease control due to better spray coverage, and increase fruit quality and size. Topping these trees would still be beneficial in the long run since it would help them regain their skirt areas and bring them to a more manageable height (Reddy and Patil, 2004; Srivastava, 2007). Optimum tree height depends on the distance between trees, hedging angle and tree width. Lower heights are sometimes used for training trees, increasing fruit size or rejuvenating declining trees. Taller trees are generally topped lower because the more limited root systems will usually not support as much top growth and may develop die-back problem. If heavy cutting is required, in older trees, reiterative pruning in phase manner is required (Gonzalez *et al.*, 1996; Reddy and Patil, 2004) .

Closely-spaced orchard /HDP : Pruning vigorous trees in a closely-spaced planting can lead to a perpetual problem of excessive vegetative growth at the expense of fruit production unless properly carried out with maximum care and precision. Severe cutting should be avoided as much as possible in case of HDP, if delayed. The more vigorous the trees and the closer the spacing, the sooner pruning should be started and the more frequently it should be done. This may be avoided by more frequent hedging of less mature wood or current season growth (Phillips, 1978).

Skirt Pruning : Pruning to raise tree skirts has become a more widely accepted practice in case of citrus having tall tree habit of growth. Until relatively recently little such pruning is done because of possible yield reduction and added expense (Phillips, 1978; Singh *et al.*, 2006). With low tree skirts, the movement of herbicide booms and other equipment is impeded, and the inspection of irrigation systems and conduct of inter-cultural operation beneath the canopy is more difficult. Fruits and limbs near the ground are often damaged by the passage of such equipment by herbicide spray and fertilizer contact. Low tree skirts may also increase the incidence of *Phytophthora parasitica*, the causal agent of foot rot, because of poor air circulation under the tree canopy. Lower canopy fruit is also more susceptible to brown rot, the result of *Phytophthora citrophthora* infection under certain environmental conditions.

Pruning after Freeze Damage : Corrective pruning should be delayed until the full extent of freeze damage can be determined. Sufficient time should be allowed for new growth to take place and for dieback to cease. Premature pruning can result in removal of some sound wood and not removing some which will continue to die back, both of which can slow tree recovery. Therefore, pruning of mature trees should not be done for a least 6 months after severe cold damage has occurred. All pruning cuts should be made into living wood below all serious bark damage and at crotches, or to a point where vigorous new shoots are growing. Further, light pruning may subsequently be done when needed to aid in forming a new framework (Wheaton *et al.*, 1986).

Reiterative pruning : Loss of tree vigor in older trees results in falling of the foliage and low yields of small fruits. It is usually accompanied by dieback of twigs and small branches. This decline may be due to age, disease, soil pests, water-induced root damage etc. The cause of decline should be determined and corrected if possible; otherwise, response to rejuvenation pruning will be temporary. If the trunk and basic scaffold limbs of trees are not structurally sound due to disease, heart rot, and cold cankers, such rejuvenation is not justified. A limited root system due to a high water table, a hardpan or poor subsoil can result in early tree decline with trees being more readily stressed by drought. When this occurs, the top becomes out of balance with the more limited root system with consequent insufficient uptake of water and nutrients by the roots. Topping these trees should temporarily restore them to a more favourable top to root ratio, but not alleviate the long-term problem. The severity of pruning for rejuvenation depends on the cause and degree of decline. Skeletonization, the removal of all foliage and wood smaller than half inch in diameter invigorates the tree and results in production of fruiting wood throughout the remainder of the tree.

The declining citrus orchards can be rejuvenated by adopting recommended pruning practices skillfully in steps. The steps are :

- In tall trees mark the height at 2m from the ground level.
- Remove the top branches bearing leaves and twigs to skeletonizing the tree.
- Carefully remove the dead and dried wood with sharp instruments before going for final pruning operation.
- Topping off up to desired marked level is done by reiterative pruning in phase manner.
- Immediately after pruning, treat all cut surfaces with Bordeaux paint. The entire body of the trunk/limbs should be brushed with Bordeaux paste.
- Thereafter cleaning of the orchard floor from the debris / cut wood logs/ branches, twigs-leaves for maintenance of phyto-sanitation.
- Removed wood portion should be sold off for the use of fuel wood only.
- Follow of the agronomic management practices as an adult tree phase.

Precautions

Application of manures and fertilizers through ring method and irrigation should be followed only after 20-25 days of pruning operation otherwise sudden collapse of the tree physiology may lead to death of the treated tree. Certain virus diseases can be spread by pruning equipment, particularly hand-held equipment. Caution should, therefore, be exercised when moving equipment from a suspected virus infected block to one assumed to be clean. As with any orchard equipment, good sanitation procedures reduce the incidence of pest and disease movement.

Nutrition

Since, pruning results a significant reduction of the mature tree canopy (15% to 100%), treated trees requires externally applied amount of especially nitrogenous and potassic fertilizers. In this way, excessive vigour of re-growth is reduced. Improved fruit set may reduce the number and intensity of subsequent leaf flushes, and facilitate pest and disease management. Judicious use of manures and fertilizers helps to maintain productivity of citrus orchards. Right choice of fertilizers in an appropriate dose in split applications at critical growth stages of the crop results into maximum advantage. The fertilizer doses recommended are on higher side. The desirable fertilizer dose provides maximum yield without affecting health of the trees. Orchards should be green manured after every 2-3 years. The fertilization should be adjusted on the basis of soil and leaf analysis. Leaf analysis can indicate the quality of the effective absorption of nutrients and their distribution to the different plant parts. Rejuvenated orchards should be fed properly as an adult dose by proper recommendation or as per schedule depending upon the situation. A dose of 50 kg well rotten FYM, 2 kg Neem cake, 1.50 kg Urea, 2.00 kg Single Super Phosphate and 1.0 kg Muriate of Potash should be applied preferably in two split doses, one during the end of May-June and remaining half dose in September (Singh *et al.*, 2004; Srivastava and Singh 2004; Srivastava and Singh, 2007).

Irrigation

Irrigation rationalization is must to restore the growth and development physiology of heavily pruned trees (Singh *et al.*, 2004).

Mulching

Black polythene mulch has been found effective for minimizing the weed density in basin area of Khasi mandarin, while *Crotolaria tetragona* and *Tefrosia candida* mulches have been found better for plant growth. Pruning large trees results in literally piles of pruning debris, a rich source of carbon many other elements (Calabrese and Panno, 1992), which are recyclable if soils are correctly managed, results in favourable changes to soil physical, chemical and biological properties. These effects include reduced water requirements of citrus trees, reduced soil erosion, compaction, crusting and cracking. It also reduces weed growth, darkening of soils and increased soil aggregation, reduce soil splash onto low-hanging fruits. The increased activity of soil macrofauna (earthworms) and microflora (saprophytic fungi) and the rapid breakdown of pruning debris and fallen leaf litter, through feeding activity of the soil food web. This breakdown is most rapid in subhumid or arid climates where trees are irrigated with under-tree micro-sprinklers. Finally, differences were observed in citrus root density, gross morphology, and distribution within the soil. These are significant factors affecting production. Trees with such root systems are more buffered against various environmental stresses caused by nutritional or drought factors, and others (Neumann and Laing, 2006). Finally, to optimise production and management efficiency, changes need to be made to such practices as irrigation, fertiliser applications and weed management where mulches are applied. Growers who mulch trees may have a powerful tool at their disposal for reduction of inoculum levels.

Intercropping

There is space between rows created after hard core pruning for rejuvenation. The arrangement for all inputs and irrigation for intercrop and main crop should be independent. Another important considerations that avoid exhaustive nature of crops as intercrops in rejuvenated orchards and also intercropping in commercial bearing orchards. Intercropping of leguminous crops, i.e French bean, soybean, blackgram, bean, cowpea, with Khasi mandarin has been standardized in NEH region. Indiscriminate growing of intercrops like potato should be avoided even in rejuvenated orchards. Leguminous crops like peas, green gram and gram should be grown leaving the plant basins undisturbed. If wheat is to be sown, separate irrigation system should be prepared for the plants to provide irrigation during March-April when wheat doesn't require irrigation. Chillies, Barseem and Jowar/Maize should not be grown in pruned / rejuvenated citrus orchards.

Management of Physiological Disorders

Fruit cracking or splitting

Cracking or splitting is common physiological disorder of citrus fruits which is especially related with limes and lemons. Lemons are more prone to fruit cracking than limes. Splitting may be radial (longitudinal) or transverse, radial being more common. The split may be short and shallow or it may be deep and wide, exposing the segments of the juice vesicles. The cracked fruits are breeding ground for micro organisms making unfit for human consumption.

Among evergreen fruits, the citrus group of plants have also been found to respond better when pruned. In growing habit, most of the citrus species/cultivars are upright growing. However, the terminal part of the upright branches has marked effect of dominance on its lower buds, which fail to sprout and give lateral.

Puffing

Puffing is a physiological disorder of citrus fruits, when it reaches to maturity. The causes of this disorder, has been related to the water exchange regulation through peel (Kawase *et al.*, 1981). Accordingly, high values of RH together with high temperatures at fruit colour break increase the appearance and intensity of puffing, particularly a period of drought.

Management of Insects and Pests

Citrus psylla (*Diaphorina citri*)

It remains active throughout the year with peak period of infestation from March to November. Both nymphs and adults suck the cell sap from leaves, buds and young shoots resulting into loss of plant vigour. The terminal shoots may wilt and dry, plant growth is affected. It also acts as a vector of greening disease, which is responsible for citrus decline. The adult psyllid is a wood coloured and can fly actively. They also inject toxins alongwith saliva which makes the leaf buds, flower buds and leaves to wilt. It rests on the leaves with closed wings and hind end raised upward.

Spray Rogor 30 EC (Dimethoate) @ 2.5 ml/l or Nuvacron (Monocrotophos) @ 2 ml/l of water or Imidacloropid @ 5 ml/l. In severe infestation, apply spray during February-March and July August. Avoid spraying if already sprayed for leaf miner. Pest population may be reduced considerably by conserving natural enemies such as Lady bird beetle.

Citrus leaf miner (*Phyllocnicnistis citrella*)

It is a serious pest of young growing plants and the nursery. It affects particularly at the time of new sprouts coming out from the pruned (rejuvenated) trees. Young leaves with zigzag serpentine mines and causes distortion and drying up of the leaves. The photosynthetic activity of the foliage is greatly reduced, which adversely affect the vigour of the plants. Attack by the pest also encourages the development of citrus canker. Growth and yield are reduced in older plants.

Prune and burn the infected leaves during inactive growth period.Apply 3 % Neem Seed Kernal Extract (NSKE) just before and after flushing. Conserve natural enemies such as Green Lace Wing, Mantids, and Spiders. Spray Monocrotophos @ 1ml/l of water during flushing period. Change the pesticide for each flush for getting good response from the insecticides.

Citrus trunk-borer (*Anoplophora versteegi*)

Sawdust like powder on the trunks usually near the ground indicates the presence of borer. The grub bore into the sapwood at ground level and tunnels upward. Affected trunk and branches dry up leading to citrus decline.

Swab the main stem/trunk with petro or any other fumigants or inject 5 ml Monocrotophos / live bore hole and plug (seal) with wet mud.

Citrus caterpillar (*Papilio demoleus*)

Young larvae feed on fresh leaves and shoots. Later on, they feed on older leaves and entire plant may be defoliated. Heavily attacked plants bear no fruits.

Hand pick various stages of the pest. Spray 3% NSKE just before and after flushing. Conserve natural enemies such as wasps.

Soft green scales (*Cocus viridis*)

Nymphs and adults suck the sap from foliage, twigs and stems, resulting in velvitalization of plants and fruits. Leaves covered with sooty mould. Remove and destroy the infested plant parts. Spray the fungal pathogen (*Verticillium lacanii*). Spray Monocrotophos 1ml /l after pest infestation/incidence.

Aphid (*Toxoptera, gossypii* spp.)

Aphids attack citrus during December-March. It sucks the cell sap from tender shoots and leaves. The growth is adversely affected. Spray Nuvacron (Monocrotophos) @ 2 ml/l or Rogor (Dimathoate) @ 2.5 ml/l or Metasystox @ 2 ml/l of water.

Lemon Butterfly (*Papilio detnoleus*)

It is highly destructive pest of all citrus species. The beautifully coloured larvae are seen on young leaves on terminal shoots. It eats up the leaf lamina margins towards mid-rib. Later on, it may feed on mature leaves and defoliate the whole shoot. The naked shoots on the tree are easily identified by looking at the faeces on the ground.

The best method is to kill the larvae manually as it is easy to pick up the caterpillars from the young plantations and nursery plants. Incase of severe infestation spray Dursban 20 EC Chloropyriphos @ 3-4 ml/l or Ekalux 25 EC (Quinalphos) @ 2-3 ml/l or Thiodan 35 EC (Endosulphan) 2 ml/l of water or Sevin 50 WP (Carbaryl) @ 2-3 g/l of water.

Citrus white fly (*Dialeurodes citri*) and **Citrus black fly** (*Aleurocanthus woglumi*)

Both nymphs and adults cause the damage by sucking cell sap from the tender foliage. The attacked leaves curl back ward and drop. Sooty mould develops on the honeydew excreted by the nymphs. The black coating interferes with the photosynthetic activity of the plants. Even fruits show black coating of sooty mould.

Spray Thiodan 35 EC (Endosulphan) @ 2 ml/l or Nuvacron 36 SL (Monocrotophos) @ 2 ml/l of water in April and September. Avoid close planting and over irrigation.

Mites (*Eutetranychus orientalis*)

Mites become serious pest after rainy season. Leaves, flowers and fruits are damaged. The leaves have minute specks left by the feeding of mites. Leaves generally give dusty looks. In severe infestation, leaves show mottling and finally drop. Skin of fruits show unpleasant patches. Marketability of fruits is greatly reduced. Spray Rogor 30 EC (Dimethoate) @ 2ml/l as soon as attack is seen in May-June.

Leaf folder (*Psorosticha zizyphi*)

Both young and mature plantations are attacked from May to October. The larvae feed inside the leaves by webbing them together and start feeding from top to downwards. The plants remain stunted.

Spray Nuvacron 36 SL (Monocrotophos) @ 1.5-2.0 ml/l of water or Dursban 20 EC (Chloropyriphos) @ 2.5 ml/l or Ekalux 25 EC (Quinalphos) @ 2 ml/l.

Bark eating caterpillar (*Indarbela quadrinotata*)

This is a pest of neglected orchards. Caterpillars eat by making holes in the wood, particularly near scaffold croches. It feeds on bark under the cover which consists of fine pieces of wood and pellets of excreta.

Keep the orchard clean. Remove the webbing from the holes with strong wire and inject Chloropyriphos 50:50 solution or kerosene oil and plug the holes with cotton.

Management of Diseases

Phytophthora disease

Among the soil borne pathogens, *Phytophthora* spp. are the major problems of citrus industry. It is a highly destructive plant pathogen, causing foot rot, root rot, crown rot, gummosis, leaf fall and brown rot diseases. Citrus orchards, which are infected/ infested with this pathogen are in declining phase. The species responsible is *P. parasitica.* This is mainly responsible for causing low productivity and short life span of citrus (Khasi mandarin in NEH regions).

The first indication of infection is the appearance of gum drops on the bark surface of the trunk usually below or just above the bud union. Infected bark and wood becomes brown to dark brown. The bark develops vertical cracks. The infection spreads rapidly to the upper trunk and can also appear on branches/ limbs. When gumming starts close to the soil, the disease spreads on the main roots and there is girdling of the bark of the trunk as a result of which the tree starts declining. Prior to death of the tree usually blossoms heavily and dies before the fruit matures.

At the time of planting keep bud union well above (30 cm) the soil level. Avoid injuries to trunk and crown roots. Do not pile soil around the trunk and provide proper drainage condition. Use of copper fungicides and some systemic fungicides viz Ridomil or Dimethomorph is effective against this fungus.

Die-back / anthracnose

The twig and foliar fungal diseases are important diseases occurring in entire citrus-growing areas. Die back and Twig blight appears generally on weak trees in the neglected orchards through out the country. Sweet oranges, mandarins especially kinnow, limes and grapefruits are highly susceptible. A number of stress factors like nutritional deficiencies, drought, attack of insect pests, virus and virus-like diseases and root rot infection by *Phytophthora* spp. together contribute to the problem. High humidity and temperature around 25^{0}C are optimum for its development of the disease under field conditions. The plants affected with one or more of the above factors show drying of twigs starting from the tip and die back. In wet weather, during and after rains, *Colletotrichum gloeosporioides*, *Botryodiplodia theobromae* and in some cases *Fusarium* spp., multiply on the dead tissue of twigs. Under stressed condition of plant, *Colletotrichum gloeosporioides* and *Botryodiplodia theobromae* together increase the intensity of mandarin twig blight gradually from season to season. After 2-3 years, the twig blight becomes prominent. Reduction of canopy volume due to this malady reduces the yield considerably.

Remove predisposing factors responsible for weakening the plant vigour. Regular pruning of dead twigs 1-2 cm below the dead portion after harvesting (during February) and destroy or burn to reduce the primary source of inoculums. Spray with Bordeaux mixture (2:2:250) or copper oxychloride 50 WP (0.3%) or Benzimidazole fungicides twice at monthly interval after pruning keep the problem under control (preferably in March and again during July and September).

Powdery mildew

Powdery mildew is caused by *Acrosporium tingitaninum* (Syn. *Oidium tingitaninum*). The pathogen attacks all aerial parts of plants in nurseries and orchards. White powdery patches of mildew appear on upper part of young leaves and on twigs. The leaf tissue at infection court in initial stages of infection turns darker watery green than normal and later becomes yellow. In severe conditions, infected leaves and premature fruits drop off and twigs show die back symptoms. High humidity and cloudy weather favour the disease development and spread. The fungus overwinters as dormant mycelium in the buds and with the onset of favourable conditions it multiplies and causes the disease. The attack of powdery mildew drastically reduces the vigour and yield. Since the pathogen attacks all the citrus cultivars and no citrus cultivar is thought to be immune to the disease, regular monitoring of nurseries and orchards is required after summer rains during predisposing climate to note first appearance of the disease.

Water shoots should be pruned regularly which are supposed to catch early infection. The disease can effectively be controlled by spraying of Tridemorph (Trade name Calixin @ 0.5 ml/l), Triadimefon, and Dinocap and Benzimidazole fungicides at first sign of attack and should be repeated at 10-day interval. Dusting of sulfur at 8-day interval is also recommended to control the disease effectively.

Citrus scab (*Elsinoe fawcctti*)

Scab becomes serious on many commercial citrus cultivars during period of high humidity and rainy season. Orange, lime, lemon, grapefruits, mandarins including kinnow are found to be susceptible in varying extent. Scab is characterized by irregular protuberant corky outgrowths up to 3mm across and deeply cracked on the invaded side of leaf, and a corresponding depression on opposite side. Infection on fruit deteriorates their quality. The disease is caused by *Elsinoe fawcettii*. When surface is wet and temperature is around 16-23^{0}C, the fungus can readily invade the young leaves, tender twigs and developing fruit to establish infection.

Do not use infected nursery plants. Old infected leaves and twigs should be removed from the orchard to minimize the inoculums load and re-infection. The disease coincides with the emergence of new flush, prophylactic sprays of Captafol, Benomyl, Methyl-thiophenate and Carbendazim at this time provide satisfactory control. Copper fungicides, if applied as protectants after bloom give good control if applied from last week of June to August at 20 days interval.

Citrus Canker (*Xanthomonas citri*)

Canker is most devastating diseases of citrus. It assumes serious proportions after the rainy season. Disease appear on leaves, twigs and fruits. On leaves yellowish spots develop. Which gradually enlarge, turn rough and brownish, and become raised on both sides of leaves. These spots are surrounded by a yellow halo. On peel of the fruit, the lesions become rough and corky. Kagji lime, lemon, oranges, citrange and grape fruits are highly susceptible. Sweet orange, baramasi lemon and mandarins are less succeptible. The spots on the twigs are more prominent as irregular and corky brown lesions. The lesions may coalesce to form large necrotic areas which lead to drying of twigs.

High humidity, well distributed rainfall and temperature from 20^{0}C to 30^{0}C provide optimum conditions for disease development. The bacterium enters the tissue through stomata or through wounds. Affected leaves and twigs bearing old lesions serve as source of inoculums. Splashing rain, infected planting material, close spacing plantations favours the fast spread and encourages the development of disease.

Spray 100ppm of Streptocycline in February, October and December. Dissolve 10 gram Streptocycline in 100 litres of water. Add pure copper sulphate @ 2g/l. of water. Three sprays of Bordeaux mixture are also effective in the control of canker.

Gummosis (*Phytophthora palmivora*)

This disease is known by different names, viz. foot rot, root rot, collar rot or crown rot etc. Normally, orchards with poor soil drainage, or flood irrigated are prone. Fungus causes rotting of rootlets, dropping of the blighted leaves. The infected plants flower heavily. Profuse gumming from the stem and branches. Shreading of bark at bud union configuration lengthwise in strips after drying.

Collect and burry deep in soil the diseased leaves or fruits. Use Cleopatra as a rootstock. Avoid flood irrigation. Paint the trunk portions with Bordeaux paint upto 20 cms of height. Spray Bordeaux mixture 2:2:250 thrice, i.e. in February before flowering, June and end July or give two applications of Ridomil MZ as paint (2g/100ml of Linseed oil) to the infected portion of the trunk and drench (25g/101 of water/tree) the soil at the base of the tree in February-March and July-August or two, sprary of Aliette 80 WP (2.5 g/l of water) in April and September.

Wither tip (*Colletotrichum gloesporioides*)

It is also called die back (anthracnose). It is caused by a fungus *Colletotrichum gloesporioides* or may be due to physiological causes. Symptoms of anthracnose appear on leaves, young shoots, and fruits. On leaves the necrotic spots appear as acervuli arranged in concentric rings. Dead parts of twigs give silvery grey appearance. Leaves shedding and twig die back are characteristic symptoms of wither tip. In severe cases, the stem and infection of immature fruits result in fruit drop and show tree die back.

Follow the recommended practices. Prune the diseased wood, leaves and fruits and destroy. Spray the affected trees with Bavistin 1.0g/l of water in February and Bordeaux mixture 2:2:250 in March and repeat the same in July-August.

Pink disease (*Pellicularia salmonicolor* Syn. *Corticium salmonicolor. Pellicularia salmonicolor* syn. *Corticium salmonicolor.*)

The disease is very destructive in tropical high rainfall areas. This disease attacks all citrus cultivars. Infected trunk, limbs and twigs turn pink in colour due to pink colour of mycelial covering, hence the disease is named Pink disease. The pathogen also attacks surrounding plantation of rubber, cacao, coffee, mango, Jackfruit etc.

Infected and damaged plants should be identified before monsoon and the affected portions of such plants should be pruned and burnt. This operation should also be done on other collateral hosts near the orchard. Cut ends and scrapped portion of plants should be covered with Bordeaux paste.

Melanose or stem and fruit rot (*Phomopsis citri.*)

This fungus manifests as dark circular depressions and yellowish margins on leaves, branches and fruits. Later, the spots become raised rough and light brown and yellow margins disappear. Sand paper texture develops on the surface of leaves and fruits. Spray 2:2:250 Bordeaux mixture during July-August and September.

Virus and virus like diseases

Virus infection of citrus is always difficult to control and usually controlled by vector management. Citrus Tristeza virus is vectored by the black aphid. Greening disease (Huang Lung Bing) is vectored by psyllids. Both important diseases have no general cure. The following are the important but difficult diseases can be prevented/controlled after rejuvenation of declining orchards.

Greening: It is caused by bacterium and manifests as stiff upright multiple twigs and buds. The leaves become small and mottled. Premature defoliation and branches show die back. Use disease free bud wood. Control citrus psylla (*Diaphorina citri*) as it is a vector for greening disease.

Tristeza: It is caused by a virus. It causes vein clearing in young leaves. The infected trees appear as their roots are damaged. This is due to sieve tubes necrosis at the bud union and crease formation, which check carbohydrate translocation from top to roots resulting in starvation of roots. Due to heavy bearing, trees usually exhaust. Inverse pits are seen on the bark at the bud union configuration.

Use tolerant rootstock as Cleopatra and Jatti khatti. Control the insect vector (aphid) which causes its spread by following the sprays of insecticides.

Exocortis (Viroid): The bark below the bud union shreads. The wood become naked at the soil level. Roots get damaged. Trees show stunting. Karna khatta and citrangs, viz. Troyer and Carrizo are susceptible to exocortis.

Use tolerant rootstock like jatti khatti and cleopatra. It spreads through cutting tools. Sterilize the tools before use in the orchard.

Ring spot : It appears as yellow rings on mature leaves. Rings may be few to many per leaf. Rings may coalesce to form bigger patches. The severely infected plants show die-back and decline. The virus spreads through infected bud-wood. Use virus free bud wood for raising of nursery.

Citrus nematode (*Tylenchulus semipenetrans*) : The female adult is oval and remains attached to the roots and suck cell sap as a result terminal growth is infected. The general plant vigour is adversely affected. Nursery should be raised on nematode free soil.

Fruit Yield

Citrus fruits should be harvested when they are fully mature and have attained attractive skin colour (Kahlon, 2008). Fruit bearing starts after one year of re-growth taking place on the pruned branches for rejuvenation and that too of high quality grades. The fruits should be harvested when their outer surface is dry. The sweet oranges, grape fruits, limes and lemons are hand picked. The thin-skinned tangerines should be harvested with clipper leaving as short stalk as possible. Clean and size grade under shade and send the bigger fruits for fresh market and smaller fruits should go for processing for higher income and profitability (Kahlon, 2008).

Unproductive citrus tree

Dense canopy

Declining tree

Die back affected tree

Neglected tree

Decling tree in orchard

Declining plant tree

Developed branch

Dense branching

New shoots

Dense branching

Thinning of branches

Developing branches

Canopy development

Rejuvenated tree

Healthy citrus tree (rejuvenated)

Chapter 12

Guava

Guava (*Psidium guajava* L.), belonging to family Myrtaceae, is cultivated and naturalized in the tropical and subtropical regions of the world. It occupies fifth place in area (0.15 million ha.) and fourth place in fruit production (1.80 million tonnes) in India. It has gained considerable prominence owing to its high nutritive value, availability at moderate price, a pleasant aroma and good flavour. Its fruits are used for both, fresh eating and processing. It excels most other fruit crops in productivity, hardiness, adaptability and vitamin C content. Besides its high nutritive value, it bears heavy crop every year and gives handsome economic return involving very little inputs. This has prompted several Indian farmers to take up guava cultivation on a commercial scale (Chadha, 2003). Guava trees are not difficult to grow and can survive in a range of soil and climatic conditions. However, precise management is required to produce a highly profitable crop.

PRESENT STATUS

Guava is important in international trade and domestic economy of several countries in warmer climate. Today, it is found growing in more than sixty countries throughout the tropics and subtropics, including some of the Mediterranean areas (Crane and Balerdi, 2005). In India, guava is well adapted in almost all states. Acreage of 1.546 lakh hectares and annual production of 17.93 lakh tonnes show the economic encouraging potential of its commercial cultivation. Guava is one of the cherished fruit crop in India. The present acreage and annual production show its potential for commercial cultivation. Boosting up the present productivity of 7 to 11 tonnes per hectare is a challenging task for researchers as well as the extension personnel. The major producing states are Uttar Pradesh, Andhra Pradesh, Madhya Pradesh, Rajasthan, Gujarat, Maharashtra, Karnataka and Tamil Nadu. Its cultivation is not seriously affected by extremes of temperature, hot winds, scanty rainfall, saline and poor soil, water logging condition and unavailability of water, fertilizer and other inputs.

PRODUCTION CONSTRAINTS

Guava is an evergreen fruit plant and is quite hardy. As a shrub or small tree, it grows to a height of 7m, and sparsely branched. Its fruits are fleshy, variable in shape and size depending on the variety. It survives under both dry and wet conditions to the extent where many other fruit crops collapse. Drastic reduction in fruit production and quality takes place, when there is any shortfall in management practices or the crop receives poor cultural practices (Singh and Bajpai, 2013).

Guava, if left untrained or unpruned, result into huge, misshappened, unmanageable trees. The bearing area is pushed away and the interior of the plant look entirely barren. Guava untill now always planted at wider spacings. Nevertheless, the recent trend of high density planting is the result of following proper training and pruning.

Over the years, productivity of guava crop continues to be low. There are various limiting factors related to production and productivity. But what makes their task unique daunting is the declining production potential of old and dense orchards existing in abundance. It has been observed that after 15-20 years of age orchards generally begin to lose vigour and bearing potential and dreaded disease of wilt accelerates their process. The essentiality of developing and deploying appropriate technology to manage such senile orchards in order to attain the competitive edge in commercial production and to meet the quality standards of the conscious consumers led to initiation of research efforts in this area. A technology to rejuvenate and restore the production potential of old, unproductive and wilt affected orchards have been developed by Central Institute for Subtropical Horticulture, Lucknow has been widely adopted and revalidated too with encouraging results and beneficial outcome. This technology employs pruning of branches at different periodicity and severities. The technology also helps in maintaining the manageable tree height with open architecture and canopy of healthy shoots with outwardly growth facilitating penetration and utilization of light. Crowding and encroachment of guava trees with subsequent inefficient light utilization, is an obvious problem with older orchards, if trees are not well managed. The internal bearing capacity of guava trees also decreases with time and age, due to overshadowing of internal bearing wood (Singh and Bajpai, 2013).

Like many other fruit crops, guava trees also witness decline in productivity as age advances, hence need even corrective pruning and to be rejuvenated, which respond excellently. The commonly occurring senile orchards, is one of the prominent factors relating to the declining trends in production and productivity of guava.

Need of Rejuvenation

The present concern is the declining production potential of old and dense orchards existing in abundance. It has been observed that after 15-20 years of age, orchards generally begin to lose vigour and bearing potential and the dreaded diseases of wilt accelerate the process. The essentiality of developing and deploying appropriate technology to manage to such senile orchards in order to attain the competitive edge in commercial production and to meet the quality standards of the conscious consumers led to the initiation of research efforts in this area at the Central Institute for Subtropical Horticulture, Lucknow (Singh and Bajpai, 2013).

A technology to rejuvenate and restore and restore the production potential of old, unproductive and wilt affected orchards has been developed, which employs pruning of branches at different periodicity and at different severities. The technology also helps in maintaining the manageable tree height with open architecture and canopy of healthy shoots with outwardly growth facilitating maximum light penetration and its utilization by the plant (Basu *et al.*, 2007; Jahangeer *et al.*, 2011).

REJUVENATION TECHNIQUE

Identification of old unproductive orchards

A large percentage of existing orchards remain, but have been neglected and can no longer attain full production. There is an important need and opportunity to rapidly enhance orchard production through the utilization of appropriate horticultural techniques. The usual survey and interactions with growers is helpful in identifying the old, senile unproductive orchards for implementation of rejuvenation technology.

Training to the growers

There is an immediate need for the development of a rapid, systematic and extensive programme to rejuvenate the orchards. Participatory survey work should be conducted to formulate an approach that can be adequately applied and disseminated for effective orchard rejuvenation. This program can be achieved rapidly and effectively through the development of training materials and the education of a cadre of trainers who will demonstrate rescue techniques to local lead farmers. There is a coincident requirement for the provision of critical inputs including pruning equipment, pest and disease control materials, budding materials etc. Subsequent activities would also include the development of nurseries and education on long term orchard management practices. Some of the elements of orchard renewal, to be integrated into any training / demonstration program, can include: Cultural practices, fertilizer management, plant protection management etc.

Pruning for Rejuvenation

During initial years, if the proper pruning or training is not carried out, its orchards become unproductive, and starts declining. These orchards need severe pruning by way of heading back its main limbs-or-sublimbs. The rejuvenation technology involved the heading back or topping off exhausted tree (showing marked decline in annual production) to the extent of 0.75m to 1.50m height above the ground level in May with facilitating production of new shoots from below the cut point and allow the development of fresh canopy healthy shoots. The newly emerging shoots are allowed to grow up to a length of 40 to 50 cm, which could be attained in 4 to 5 months of pruning. These shoots are further pruned out about 50 per cent of its total length in October to facilitate emergence of multiple shoots below the pruning point. This was mainly done to modify the tree structure and maintain canopy size. Profusely emerging shoots in the inner canopy are also pruned out to promote branching. The multiple shoots developed as a result in October pruning are capable of producing flower buds for rainy season crop. The farmers keen to take rainy season crop can allow the shoots to bear buds and fruits (Kalloo *et al.*, 2005).

However, as winter crop has more marketing edge and value due to quality, taste and fruit being free of pest incidence, it is desired to promote fruit load in winter season. Hence to check the onset of rainy season crop, shoot pruning (50%) is also done again in May. Emergence of new shoots is facilitated. These new shoots emerging after May pruning are found to have high flowering and fruiting potential for winter crop. This procedure of sequential and periodic pruning is continued every year for proper shaping of tree canopy and to ensure enhanced production of quality fruits during winter season. The cost of rejuvenation technology and the loss of yield in the first year are compensated with sale of wood procured from pruned trees as well as the enhanced income generated by raising intercrop in rejuvenated orchard (Singh and Chanana, 2005).

In totally declined plants it is advised that they should be headed back to the extent of 0.75m-1.50m above the ground level in May to allow the development of fresh canopy of healthy shoots.

These newly emerging shoots are allowed to grow up to a length of about 40-50cm. These shoots were further pruned to 50% of its total length in October for emergence of multiple shoots. It is desired to promote fruit load in winter season. The emergence of new shoots is facilitated. These new shoots emerging after May pruning have high flowering and fruiting potential for winter crop. This procedure of sequential and periodic pruning was continued every year for proper shaping of tree canopy and to ensure enhanced production of quality fruits during winter season (Singh *et al.*, 2005).

Canopy development: Pruning at definite periodicity as standardized, rejuvenated the trees with attainment of healthy and vigorous canopy and development of short tree architecture conducive for canopy management, plant protection and cultural and harvesting operations (Singh *et al.*, 2007). Fruit load with better quality also shifted significantly towards the interior of canopy. New growth and inducement of bearing of shoots are stimulated in the inner canopy in the same way as the hedging and topping encourages new growth on the outside of the canopy. Tiller like growth of undesirable shoots and branches need to be removed for judicious use of nutrition and development of productive canopy. Pruning at definite periodicity as standardized above rejuvenated the trees with attainment of healthy and vigorous canopy and development of short tree architecture conducive for canopy management, plant protection measures, cultural operations and harvesting operations. Grower should follow the cultural practices prudently for survival of trees and successful utilization technique.

Upright branches generally remain vegetative and vigorous. Horizontal branches are generally more fruitful. A good combination of these two is necessary for fruiting in subsequent seasons. Therefore horizontal growth of branches should be promoted near upright growth to facilitate maximum bud growth. The level of net photosynthesis was more pronounced in flowering branches of tree which was topped from 1.5 m height as compared to the non-flowering ones in both cultivars. Light penetration was greater in 1.5 and 2.0 m height pruned trees internal relative humidity of leaf was highest under the control (no pruning) and decreased with the increase in pruning intensity. Canopy temperature was maximum in rejuvenated orchard and lowest in control. Fruit yield increased to the tune of 75 to 80 per cent in developed canopy of rejuvenated orchard trees over the control (Singh *et al.*, 2007).

Manipulation of vegetative growth and fruiting pattern

One of the critical characteristics of guava is that flowers are borne on newly emerging lateral shoots, irrespective of time of year. Consequently, the occurrence of bloom and fruiting in the course of the year may be erratic or seasonal, depending on how the environment affects shoot growth (Salazar *et al.,* 2006. The characteristics allows the tree to be manipulated to crop when desired in a favourable climate. Pruning is the main method ot force the axillary bud to shoot. Presumably, the flowers are already differentiated before the side shoots emerged, implying that the lateral bud should not be forced to break before differentiation has been completed. Shoot growth is indeterminate under good growing conditions long vigourous shoots dominate, which suppress the emergence of flowering side shoots. Canopy design and shape influence light interception and higher monetary returns can be assured to guava growers. The crop yield is harvested during December-January, Arpil–May and August-September respectively. Studies indicate that overlapping of fruit growth developed as a result of pruning, makes it possible to harvest three crops in a year with proper management of new shoots and flower initiation, the young shoots can be forced again to flower and fruit in the next season. The success of this technique depends upon the proper management of shoots through precise and timely pruning. Based on the appearance of the leaf and shoots due to concurrent pruning, shoot development can be judged, which is required for better fruiting in different season.

Rejuvenating wilt affected guava trees

Wilt is a pernicious disease and a curse to guava industry. It causes enormous loss to guava plantation every year (Singh and Bajpai, 2013). Browning characteristically identifies its incidence and wilting of leaves from tip, discolouration of stem, bark splitting followed by complete mortality of branches and then whole tree. An alternative period of heading back during December-February also standardized, where in the orchard particularly those infested with wilt disease were headed back during the stated period. Experiments at CISH showed encouraging results in beginning healthy canopy as well as crop bearing potential in wilt affected trees by way of severe pruning and intensive cultural and nutrient management.

Under experimental trial wilt affected trees were headed back from 1.0 to 1.5m above the ground level during December to February. The topping height i.e. the point of heading back is generally decided on the basic of strength of disease infection and its severity as well as the tree architecture. Pruning is undertaken from 5.0cm below the infection point. It is identified through presence of green wood and healthy pith. Profuse emergence of shoots was observed on the wilt affected trees after pruning and they were allowed to grow. After 5 month of growth, i.e. May – June, the shoots were subjected to pruning to 50 per cent of their length for initiation of new shoots to bear floral bud for winter season crop. With this technique not only new canopy was restored but also yield of fruits 34.00 kg/tree was obtained.

Microbial analysis of soil : Initial results of investigations about the comparative loads of causal pathogen of wilt i.e. Fusarium oxiporum and Fusarium solani in the rhizosphere of the wilt affected trees subjected to pruning for rejuvenation

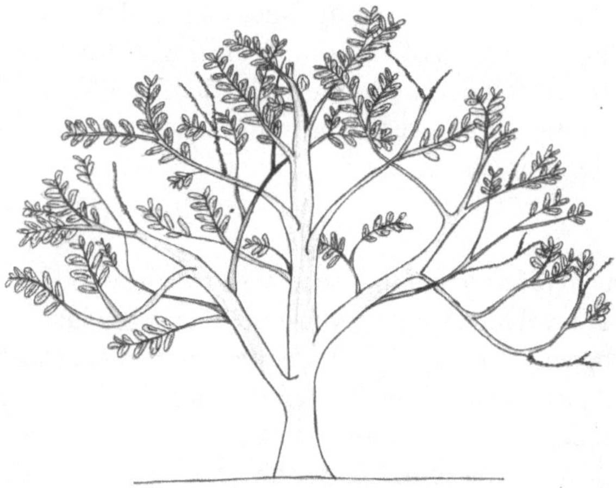

Old senile unproductive guava tree

Pruned tree

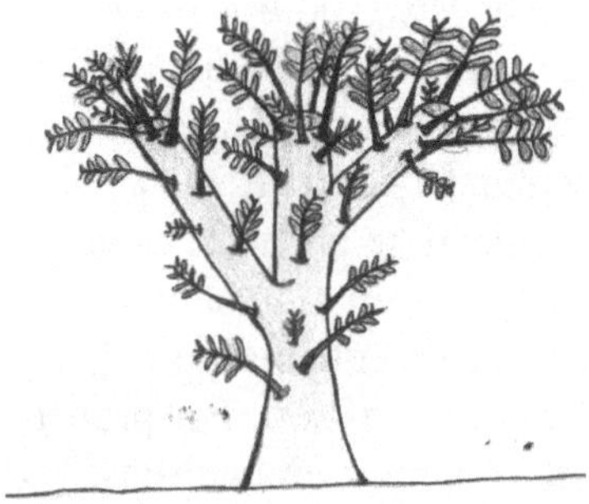

Emergence of shoots

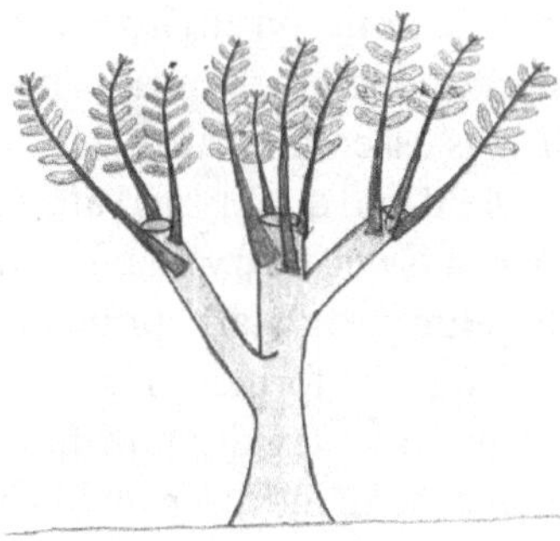

Shoot thinning

Rejuvenated tree in full bearing

of wilt affected control trees have been encouraging. Microbial analysis of soil samples showed comparatively lower load of Fusariumin the rhizosphere of the pruned trees. Observing the encouraging results of the technique in rejuvenating the senile orchards, demonstration trials were planned to study the on-farm performance in different farming situations.

Nutrition

Integrated nutrient management refers to maintenance of soil fertility and plant nutrient supply to an optimum level for sustaining the desired crop productivity through optimization of the benefits from all possible sources of plant nutrients in an integrated manner (Singh and Bajpai, 2013). Therefore, it is holistic approach where we first know what exactly is required by plants for optimum level of production, in what different forms these nutrients can be applied in soil, at what different timings in best possible method, and how best these forms can be integrated to obtain higher productivity with efficiency of economically acceptable limits on environment friendly way. Addition of inorganic fertilizer constitutes one of the most expansive inputs in agriculture. In the country like India, it is essential that such approaches are followed which do not adversely affect yield and quality and are simultaneously cost effective and eco-friendly. As such, it is necessary that fertilizer applications are made using both inorganic and organic particularly pruned orchards for initiation of new shoots on which flower buds are formed. The amount of fertilizer to be applied, depends on the age of the tree, condtion of the tree and type of soil. For proper growth and profitable yield, fertilizer should be in the required optimum dosages. The results on the effect of manures and fertilizer on the yield and growth have clearly shown that guava is very responsive to the application of inorganic fertilizers along with organic manures. Soil type, nutrient status and leaf analysis can give better indication for requirement of nutrients. Since, leaf is the principal site of metabolic activity in the plant and changes occurring in the leaf metabolic activity are reflected in the plant performance, emphasis is now being given to adopt leaf analysis as a tool to assess the nutrient needs of guava plants. Critical limits of N,P,K, Ca and Mg have been worked out, which are 1.4-1.96, 0.20-0.40; 1.31-1.70; 0.67-0.83 and 0.25-0.68 per cent respectively.

Experiments conducted on standardization of NPK requirements have identified that this depends on many factors, viz., tree age, location, climate soil type, cropping pattern and intensity. In guava a dose of 50 kg well rotten FYM, 2 kg Neem cake, 1.50kg Urea, 2.00 kg Single Super Phosphate and 1.0 kg Muriate of Potash should be applied preferably in two split doses one during the end of May-June and remaining half dose in September (Singh and Mishra 2007). In earlier recommendation during the phase of heading back, 50 kg FYM along with 6 kg neem cake / plant is made. Six month after rejuvenation, manures 40 kg FYM + 4 kg neem cake + 1300 g urea +500 g muriate of potash and 1800 g single super phosphate / plant /year to be given (Kalloo *et al.*, 2005). Deficiencies of Zinc and Boron have become widespread in guava growing area and major cause of decline, hence after reiterative pruning soil application of zinc sulphate @800g/tree, 10-15 days before flowering (in winter as well as in rainy season flowering) should be done. Two pre-blossom sprays of 0.3 to 0.5 per cent zinc sulphate at 15 days interval should be done. To avoid boron deficiency pre-flowering spray of 0.3-0.4% boric acid should be done.

Irrigation

The chief economic consideration, which encourages growers to go for guava cultivation that this tree does not suffer much, if it is not watered during hot months. But adequate moisture is required soon after heading back of branches for proper initiation and development of shoot growth. Care of pruned trees requires watering them regularly during the dry season. Due to lack of moisture stress in the pruned trees, the emergence of new shoots as well as rejuvenation process are severely affected. Therefore, it is necessary to ensure irrigation in the rejuvenated trees at regular intervals for initiation of shoots below the cut portion. To promote the proper development of tree canopies and fruiting twigs, irrigation should be carried out at the interval of every 7-10 days in summer and at about 15-20 days interval in winter, in addition to the period of rainfall during the monsoon season (Singh and Bajpai, 2013).

Orchard floor management

Regular intercultural practices are essential for proper upkeeping of the pruned guava orchards. It improves physical condition of soil, ensuring aeration by breaking soil surface crust and removes weeds those compete for soil moisture and nutrition. In order to manage the orchard soil, two time ploughing is necessary in a year. One ploughing should be done during June and second ploughing during December. The first ploughing helps in checking runoff losses and facilitates maximum intake of water into soil. The second ploughing checks the weed growth and induces vegetative shoots.

Intercropping

Besides rejuvenating the trees this technology also offers opportunities for employment and income generation through raising intercrops in the floor space of the pruned orchards. Intercropping is intended to maximize land and space use efficiency to generate supplementation income, particularly during the phase of canopy development in pruned orchards. After rejuvenating the guava orchards, tree takes about 2-3 years to develop canopy and cover the area.

However, great care should be taken in selecting the right type of intercrops at the centre, between and within rows of pruned guava trees. Vegetables and leguminous crops can easily be taken upto three year's after rejuvenation. The crops like cowpea, bean, cabbage, cauliflower, chillies, okra, and partial shade loving plants (ginger, turmeric and elephant food yam) as intercrop in the orchard provide sustainable return from the initial stage of canopy development.

Mulching

Mulching at the base immediately surrounding the main trunk in the pruned trees is done by using black polythene sheets (100 micron=400 gauge thickness) or heavy mulching with organic material such as straw, dried grass, banana leaves. It drastically reduces weed growth. Mulching with organic materials should be applied thick enough (12 to 15cm) to prevent weed growth yet permit rain water penetration to the root area. Black polythene sheets prevent soil surface evaporation and tend to produce water under the sheets through condensation, supporting tree growth besides affording weed control. There is a considerable reduction in water application in polythene mulched orchards as compared to un-mulched ones. Reduction in water application economizes the inputs, enhancing the income per unit area.

Management of Physiological Disorders

Little leaf

The leaves become small and form a cluster due to micronutrient deficiency (mainly Zn). Intervenal chlorosis, sparsely distributed foliage, reduction in leaf size and leatheryness of leaves, reduced plant growth, die back of branches, reduction in number of flowers and cracking of fruits are other characteristic symptoms. The deficient trees do not flower and also get barren with no flush. Extreme reduction in yield with poor fruit quality are also observed. It is a serious problem in waterlogged and saline area.

Soil application of zinc sulphate @800g/tree, 10 to 15 days before flowering (in winter as well as in rainy season flowering) should be done. Two pre-blossom sprays of 0.3 to 0.5 per cent zinc sulphate at 15 days interval should be done.

Internal necrosis

Boron deficiency in guava results in internal necrosis and affected portion becomes hard. Fruit size is reduced and leaves start falling. Cracking of fruits are seen in extreme cases.

Pre-flowering spray of 0.3-0.4% boric acid should be done. Spray of borax @ 0.5% (dissolved in hot water) during July and August are beneficial for fruit quality.

Management of Pest

Fruit fly (*Bactrocera dorsalis* Hendel)

The fruit fly is the most destructive insect in production of guava, particularly during rainy season. When the fly remains uncontrolled, the amount of marketable fruit is drastically reduced. Damage occurs as the larvae hatch out from eggs oviposited beneath the surface of the ripening fruit and begin to feed on the flesh. Affected fruits turn progressively soft and mushy as the larvae begin feeding, until the fruits become waterlogged and the juice begin to drip on handling.

The traps are very useful in monitoring and control of population of fruit fly. Hanging of bottle traps containing 200 ml of water emulsion of methyl euginol (0.1%) + Imidachlorapid (1ml/l) during fruiting season (April-July) is very effective for control of fruit fly. Ten traps per hectare in orchard give satisfactory control. Traps can be fixed during morning time.

Collection and destruction of infested and fallen fruits along with maggots useful in reducing the pest population. Ploughing tree basin also helps in checking the pest population as the pupae are destroyed by being exposed to unfavourable temperature and also becomes the prey for predators. Adult fruit flies can be controlled by bait spraying of Carbaryl (0.2%) + 0.4% protein hydrolysate or molasses at pre-ovi-position time. Avoid taking rainy season crop only in orchards with history of severe fruit fly infestation.

Guava shoot-borer (*Microcolona technographa*)

Guava shoot borer is active on nursery and young plantation of guava from March to December. Larva damages the tender shoots of nursery and adult trees of guava, which result in side sprouting of vegetative buds just below the larval damaged portion produce lateral shoots and give the plants a bushy look. Secondly, infested shoots dry up which can be located from distance by presence of fine black frass on the leaves/shoot beneath the site of infestation. Peak period of infestation occurs during March to December.

Spraying of Monocrotophos (Monocil @4ml/l) or Chloropyriphos (2-4ml/l) can control the pest.

Bark-eating caterpillar (*Inderbela* spp.)

It is another serious pest of guava found all over the India. The old, shady and neglected orchard are more prone to the attack of this pest. The caterpillars bore into trunk, main stem and thick branches, and remain inside the holes during the day. The caterpillars come out in night to feed on the bark and make silken galleries inside.

Kill the caterpillars mechanically by inserting the iron spoke in shelter holes made by these borers at early stages of infestation. In case of severe infestation remove the webs and insert the swab of cotton wool soaked in 0.05% Dichlorovos or inject water emulsion of 0.05% Monocrotophos or of 0.05 %, Chloropyriphos and plug the holes with mud.

Guava fruit-borer (*Virachola isocrates*)

The guava fruit-borer has been found in the northern region of the country. The caterpillars bore the raw fruits of all size and eat the pulp of the fruits. The infested fruits usually dry up.

Collect the infested fruits with borer and destroy them to check the carry over of the pest. The adults may be managed by spraying of Carbaryl (0.1%) or Fenthoate (0.05%) or Phosalone (0.01%) at the beginning of fruiting season and before the ripening of fruits. Spraying of Carbaryl (0.2%) at the early stage of crop has been found effective in reducing the pest population.

Management of Disease

Fruit canker

The fruit canker is widely prevalent in India. It produce numerous circular to raised dark coloured corky cankerous growths on fruits. Infested fruits are deformed and give a chicken pox appearance. Infected fruits do not ripe and are not palatable.

The disease can be effectively managed by 3-4 sprays of 1% Bordeaux mixture or lime sulphur at 15 days interval.

Anthracnose

High humidity and frequent rains favour the spread and intensity of disease attack. It occurs mostly on fruits during rainy season. Most characteristics symptom include the appearance of small spores of the size of pin head, which are first observed on unripe fully grown fruits during rainy season. They are dark brown to black in colour, sunken, circular and bear numerous minute black pin head growth in the centre of the lesions. In favourable weather, several spores coalesce to form bigger lesions. The diseased portions are comparatively harder than the healthy tissues. Ripe fruits become soft. Unopened flowers and buds are also attacked and they are shed.

Pruning of diseased twigs and destruction of fallen leaves and fruits are also helpful in managing the disease. Mancozeb (Dithane M45, 0.2%) or Thiophanate (TopsinM, 0.1%) or Carbendazim (Bavistin, 0.1%) or Copper-oxychloride (3g/litre of water) on mature fruits reduced the infection.

Styler-end rot

Severe infestation occurs during rainy season which reduced the quality of fruits. The symptoms start as a circular, water – soaked lesions at the styler end and later on they reddish brown in colour.

The disease can be managed effectively by spraying of Carbendzim (Bavistin, 0.1%) or Thiophanate (Topsin M, 0.1%) at 15 days interval during fruit maturity stage. However, no spraying is done before 15 days of harvesting.

Fruit rot

The symptoms appear on mature green fruits as a water soaked lesions that develop very rapidly to affect entire fruits. This is very serious during rainy season and spoils nearly 20-25% of the fruits before harvesting (Singh and Bajpai, 2013).

Remove diseased fruits from the orchard and destroy them so that fruit flies and other insects can not land on the fungal mass to pick up spores for re-infection. Pre-harvest spraying of Cardendzim or Thiphanate (Bavistin or Topsin M, 0.5%) 15-20 days before harvesting effectively manage this disease.

Wilt (*Fusarium* sp.)

In guava, occurrence of Wilt disease is widespread across the country but supposed to be more prevalent in alkaline soils (Dwivedi *et al.*, 1999; Singh and Bajpai, 2013).

The infection can be effectively minimized by drenching the soil with Brasicol and spraying the plant with Bavistin (0.1) at an interval of 10 days at the early stage of infection. Resistant varieties such as Banarasi, Nasik and Dholka should be planted to get rid of this disease.

Fruit Yield

Yield enhancement in the range of 70 to 90 per cent over the unpruned trees was recorded after second year of pruning with topping and hedging. Vegetative to reproductive balance begins to stabilize after second year. Fruit load with better quality also shifted significantly towards the interior of the canopy. However, initial yield was recorded lower in rejuvenated plants as compared to unpruned and farmers practice. An average fruit yield in the tune of 30-40 kg/plant after one year of rejuvenated trees can be obtained and which found increasing many fold in concurrent years ahead with the advancement of age. The economic analysis revealed that B:C ratios were much higher in rejuvenated plants as shown through economic feasibility analysis in case of two commercial varieties i.e. cv. Allahabad Safeda and Sardar (Nimbalkar *et al.*, 2010; Singh, 2010; Baba *et al.*, 2011; Singh *et al.*, 2012; Mishra, 2013).

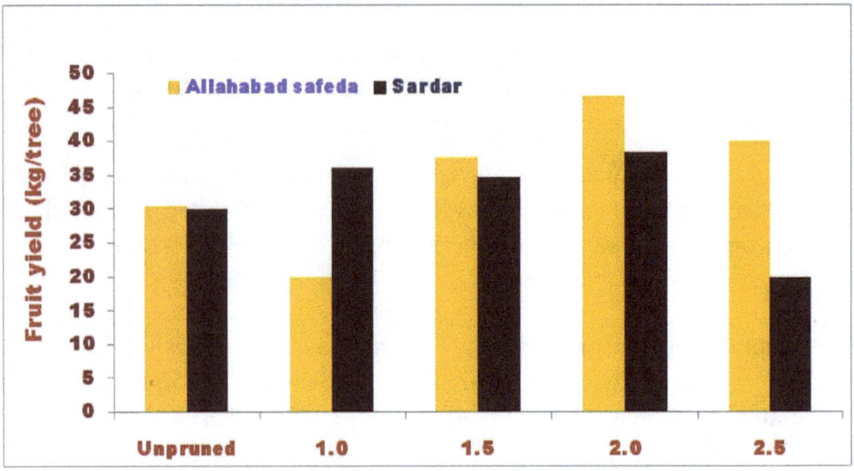

Response of pruning on fruit yield of guava cultivars

Source : Singh and Singh, 2007

FLOW CHART OF REJUVENATION TECHNOLOGY GUAVA

Identification/Selection of Old Senile Unproductive Orchards

↓

Training to the orchard grower

↓

Reiterative pruning in précised manner

↓

Right time of reiterative pruning (April-May)

↓

Application of manures and fertilizers (June)

↓

Thinning, training and canopy development

↓

Irrigation and mulching

↓

Intercropping

↓

Plant protection

↓

Fruit yield and quality

↓

Economics

↓

Impact

Old senile orchard

Declining tree

Unproductive tree

Densely overlap branched in the tree

Wilt infested tree

Tree with lichen infestation

Identified old tree

Pruning of tree

Pruning with chain saw

Pruning unwanted branches

Pruning for shaping

Heavly pruned orchard

New sprouts on pruned trunk

Shoots on pruned trunk

Invigorated shoots on single trunk

Profuse sprouting on multi trunk tree

Developing canopy after shoot thinning

Shoot thinning on single trunk

Shoot thinning on multi-branched trunk

Canopy on multi-branch trunk

Canopy on single trunk

Shoot emergence pattern

Shoots for secondary pruning stage

Productive branches

Productive shoots

Secondary stage pruning

Branch frame work

Rejuvenated trees at bearing stage

Quality fruits on rejuvenated tree

Chapter 13

Jackfruit

Jackfruit (*Artocarpus heterophyllus* Lam.), belonging to the family Moraceae, is indigenous to the rain forests of the Western Ghats of India (Menon and Peter, 2011; Amma anbd Kumaran, 2011; Alila and Sanyal, 2011). It spread early onto other parts of India, south-east Asia, the East Indies and the Philippines. It is often planted in central and eastern Africa and is fairly popular in Brazil and Surinam (Menon, 2011). Its tree is adapted to humid tropical and near-tropical climates. The jackfruit tree is handsome and stately. In the tropics, it grows to an enormous size. One of the popular cultivar, Ceylon, was introduced into India from Ceylon and planted extensively in our country.

PRESENT STATUS

India is the second largest producer of jack fruit in the world and better known as the motherland of jackfruit (Verhji and Coronel, 1992; Chadha, 2003; Amma and Kumaran, 2011). Though, it is indigenous to India, but also found widely grown in Bangladesh, Burma, Sri lanka, Malaysia, Indonesia, the Philippines, Brazil and other tropical countries (Milan, 2011). In India, it has wide distribution in Asam, Tripura, Bihar, Uttar Pradesh, foothills of the Himalayas and Kerala, Tamil Nadu and Karnataka. Jackfruit is the national fruit of Bangladesh. It is one of the three auspicious fruits of Tamil Nadu along with mango and banana. Jackfruit is very extensively grown all over the state of Tripura. As per the records of the Department of Horticulture and Soil Conservation, Government of Tripura, the area under jackfruit is about 9,012 ha with a total production of 2,69,404 tonnes (2012-13).

PRODUCTION CONSTRAINTS

In India, commercial cultivation of jackfruit is still at a primitive stage. It is because of development of new varieties with a slow pace as well as difficulty in procuring elite planting material (Milan, 2011). There is still lacking a reliable fruit yield data on jackfruit grown in India. Secondly, jackfruit is a highly cross-pollinated crop and due to its monoecious habit, and plants raised from seed never bear fruits like true-to- the- type of the mother plants. Majority of the existing orchards are propagated from seed, which make management and harvesting difficult. Though, trees produced from physiologically matured vegetative material bear fruits earlier than those grown from seed. The jackfruit trees are intolerant to poor drainage (Soepadmo,1992) and roots fail to grow under flooded conditions. In such a situation, its tree attains the declining phase, particularly due to inactivation of roots and die-back disease. Seedling trees start bearing from seventh to eighth year onwards while the grafted ones from third year, when a few fruits may develop. Singapore variety starts yielding from third year of planting but this type of variety is not widely available.

Jackfruit trees generally live for 100-120 years or even longer. However, fruit productivity declines with age, but the price of its timber increases as the tree gets older. To achieve the maximum productivity, its trees should be 12 years or even older. The fruit yield varies depending on variety that too is influenced by proper cultural management practices. As grafted trees bear fruits 4-5 years after planting, while trees grown from seeds start bearing fruits 6-8 years after planting. Fruit bearing is delayed and fruit quality is affected if the trees are grown at a higher elevation (Medagoda and Tennkoon, 2001). Fruit rot is a major problem in jackfruit. Dieback of the tree is also prevelent in neglected condition (Milan, 2011; Namuco, 2011).

Need of rejuvenation

As training and prunning are not reguloarly done as a routine practice in Jack fruit in seedling trees. However, fruit productivity declines with age. The senility and decline in production potential also depend on the type of management practices being received by trees (Hackett, 1985). The plants of jackfruit do not have recommended rejuvenation technology but better canopy architecture for higher productivity even in old existing orchards, a systematic pruning and training even through corrective pruning are required for proper sunlight penetration and mature limbs for flowering and fruiting. The shoots with die-back symptoms should be pruned and should be a part of sanitation (Alila and Sanyal, 2011; Amma and Kumaran, 2011).

REJUVENATION TECHNIQUE

Identification/Selection of Old Orchards

Mostly jackfruit trees are found growing neglected and in isolation. But more than 10 number of trees and a systematic unit orchards (40 and more) need recommended package of cultural practices for proper performance. Neglected orchards having more dried limbs, scarce foliage and die back symptoms should be identified for proper pruning and training operations (Amma and Kumaran, 2011).

Training to Orchard Owners/Awareness to Growers

Proper prunning and training operation for young and old trees should be clearly taught or demonstrated as onsite practical operation to increase quality production (Alila and Sanyal, 2011). In most of the cases, growers are interested only to sale the trees for timber or fuel, which is a great loss. In fact, major hitch in adoption of rejuvenation technology is that growers/farmers do not have the knowledge of the technology. The mere neglect of commercial bearing trees is the main reason for poor productivity and early decline in jackfruit. There is need to train the farmers about the improve of package of practices for commercial production.

Hard core pruning and training operations

In jackfruit, there are no general recommendations on hard core pruning for rejuvenation of old senile unproductive trees, but pruning and training operations have been found feasible due to its peculiar fruit bearing behaviour (Alila and Sanyal, 2011). Secondly, jackfruit does not respond well to indiscriminate pruning and it is not commonly practised. Regular pruning of weak, dead and diseased branches, and removal of parasitic plants at the end of the rainy season is recommended (Namuco, 2011). This prevents insect infestation and disease infection. Tree height and size may also be controlled, if desired, by pruning. Old flowering shoots should be removed after harvesting and then branches should be thinned out to increase light penetration to the inner canopy. The height of trees should be maintained at 4-5 m by periodic selective pruning. Selective pruning is also effective for equipment movement and other management operations. Trees may also be mechanically topped at 4-5 m and hedged at a 5-10° angle from the vertical (Muthulakshmi, 2003; Amma and Kumaran, 2011).

However, when older trees are not pruned and remain mere neglected without any management practices, a strong central leader usually develops which is desirable for its timber value. Nonetheless, where fruit production is the main purpose, hard core pruning carefully in reiterative manner is carried out up to the height of 3m from the ground level and similarly side- by- side the first lateral branches also accordingly should be pruned with the aim to slow upward growth and enhance the spreading of the canopy.

Old senile unproductive jackfruit tree

Pruned tree

Emergence of shoots

Thinning of shoots

Well developed rejuvenated tree in bearing fruit

The regeneration of new sprouts as per invigoration effect starts after 30-40 days, but shoot growth and elongation at slow rate. Then thinning of shoots/ branches be started after six months of when pruning operation is over. Training in jackfruit is needed to build a strong framework and to avoid weak crotches conducted on while developing the spreading canopy with reduced height, to allow unobstructed sunlight to stimulate flowering. These branches should be continuously pruned to get a reasonable trunk. Thereafter, selected branches may be allowed to remain but removal of vigorously growing upright shoots is recommended. Inner branches of developing canopy should be removed to allow more light and air within the canopy. Additional unwanted shoots are removed time to time to maintain the desired shape of plants. However, research is warranted for further improving the productivity through this highly scientific and skilful operation and must be carried out under proper guidance. In general, jackfruit does not respond well to indiscriminate pruning and it is not commonly practised. Rejuvenation may also be achieved through regular pruning of weak, dead and diseased branches. The removal of parasitic plants at the end of the rainy season is recommended. This prevents insect infestation and disease infection. Tree height and size may also be controlled, if desired, by pruning. Old flowering shoots should be removed after harvesting and then branches should be thinned out to increase light penetration to the inner canopy. The height of trees should be maintained at 4-5 m by periodic selective pruning. Selective pruning is also effective for equipment movement and other management operations. Trees may also be mechanically topped at 4-5 m and hedged at a 5-10° angle from the vertical.

The practice of fruit thinning is also recommended to prevent damage to branches as heavy fruit load breaks branches and can result in death or stunting of tree. Limiting the number of fruits per limb may also improve the quality of fruits and increase their size. However, not much research data is available on these aspects.

Nutrition

The manures and fertilizers are spread in the basin and thoroughly mixed with soil. A recommendation of 600g N, 300g P and 240g K / tree / year in commercial bearing tree and that to in split doses of before and after monsoon. Though research on nutritional requirement is scanty and specific recommendations are not available (Amma and Kumaran, 2011; Namuco, 2011).

Irrigation

The trees should be irrigated precisely to avoid flooded conditions. Waterlogged conditions may lead to high mortality of rejuvenating trees. The soil at the base of the plant should be raised to prevent water stagnation. Irrigation during dry periods is considered essential in arid regions for normal growth. The trees are sensitive to drought. In order to economise use of water, ring system may be adopted for irrigation. Alternatively, drip irrigation may be followed so as to economise water use. The frequency of irrigation depends on soil moisture condition and developing canopy growth. The frequency and quantity of irrigation depend on weather and soil moisture conditions, but jackfruit responds well to irrigation between flowering and fruiting. Hence after second year of pruning the regrowth starts bearing, requires proper soil moisture content and its management during critical periods.

Intercropping

When soil moisture is not a limiting factor, crops like okra, brinjal, chilli, tomato and gram, can be conveniently grown in open space created after hard core pruning. Particularly, when canopy starts developing and later pruned trees come into bearing, gram grown as intercrops are beneficial. These crops also improve the nitrogen status in the soil (Amma and Kumaran, 2011).

Management of Physiological Disorders

Irregular shape

The irregularly-shaped jackfruit is due to a boron deficiency in its fruits. Nutrition plays a vital role and boron is the key to improved flowering, fruiting, and internal and external fruit quality.

Management of pest

Insect and pests are not a serious problem in jackfruit but unmanaged trees/ orchards pose some problems due to infestations.

Shoot-and fruit-borer (*Diaphania* sp.)

It is a major pest of jackfruit in India and Bangladesh. The reddish brown caterpillars bore into shoots, buds and fruits. Affected shoots wilt and get dried. Bored fruits rot and drop (Amma and Kumaran, 2011; Alam, 2011).

Management : Destroy affected shoots along with caterpillars. Collect and destroy fallen fruits. Cover young developing fruits with perforated alkathene bags to prevent egg laying. Spray of Profenophos (1 ml/litre) or neem oil (10 ml/litre), and alternatively, Dimethoate (2.5) ml/litre). Trees showing infestation sign at immature stage of fruits should be sprayed with Carbaryl (0.1%) or Diazianon (0.2%) at 15 days interval for 2-3 times (Alam, 2011).

Bud weevil / leaf-eating weevil

The bud-weevil (*Ochyromera artocarpi*) is a specific pest on jackfruit. The small whitish grubs bore into tender flower buds and fruits, inducing premature fruit drop. The greyish brown adult weevils are found nibbling the leaves. Two more weevils, *Onychocnemis careyae* and *Teluropus ballardi,* and reported feeding on leaves in South India.

Removing and destroying the affected fallen shoots, buds and fruits and spraying of Carbaryl (3 g/litre) are control measures..

Aphids (*Greenidia artocarpi* Westw. and *Toxoptera aurantii* Bd.F.)

Both species of aphids feed in colonies on the underside of tender leaves and shoots which get destroyed and devitalized. Aphids can be controlled through: (a) spraying *mahua* oil (2%), (b) spraying of FPE, Nimbecidine @ 3 ml/litre of water/spray application of Biomagic-F @ 3 kg/ha and neem oil (1%) or Dimethoate (0.03%) or acephate (1.5g/litre). Do not harvest fruits up to 2 week after spraying.

Giant mealy bug (*Drosicha mangiferae* Gr.)

It is a major pest (Bhutani, 1978). Flat, oval, waxy white bugs observed in clusters on tender shoots and inflorescence cause damage by sucking vital sap. During dry weather it moves down and inhibits the roots. Prolonged drought may result in servere outbreak (Alam, 2011). For managing this pest, control measures are : (a) plough orchard in summer to expose eggs to natural enemies and sun heat, (b) remove weeds like *Clerodendron infortunatum* which are alternate hosts of giant mealy bugs, (c) apply 25 cm wide alkathene bands on tree trunk to prevent migration of crawlers from soil to trees, one week before their emergence, (d) spray crude garlic oil (1%) on tree trunk below band to kill the crawlers congregated, and (e) conserve natural enemies like *Coccinellids* and spiders by avoiding application of broad spectrum insecticides during peak activity period. The nymphs congregrated below used polythene/alkathene ponds and can be killed by spraying 0.2% carbaryl or 0.1% Cypermethrin (Alam, 2011).

Trunk-borer

A number of wood borers that attack trunks and branches are reported from Florida. These are *Elaphidion mucronatum*, *Nyssordrysina haldemani* and *Leptostylopsdis terraecolor*. In Asia, they include *Apriona germani* (Soepadmo, 1992). These bore into the trunks and branches. Affected parts show signs of wilt and subsequently dry up and severe infestation may cause fruits to drop. The borer can be controlled by : (a) collecting the grub using iron hook, and (b) fumigating with Dichlorvos and plugging the hole with mud (Alam, 2011).

The bark-eating caterpillars (*Inderbela tetraonis* Mo. and *Batocera rufomaculata* DeG.) also attack jackfruit (Azad, 2000). Jack scales is the name given to a number of different species: lesser snow scale (*Pinnaspis strachan*)i, coconut scale (*Aspidiotus destructor*), mango shield scale (*Protopulvinaria mangiferae*), and pyriform scale (*Protopulvinaria pyrifomis*). These can be found on leaves, stems and fruits. Other insect and pests affecting jackfruit include sucking insects such as mealy bugs, fruit flies and thrips (Soepadmo, 1992). Reddy (1998) reported *Crossononema malabaricium* and *Neolobocrilonema palamiensis* from soil around the roots of jackfruit in India.

Management of Disease

Pink disease (*Cortitium salmonicolor*)

It is widespread in tropical and subtropical areas. Young branches of trees lose their leaves and show die-back. Thin pink inclusions are seen on lower shaded side of the leaves.

The disease appears as a pinkish powdery coating on the stem. Pink colour represents profuse spore production of fungus (Tangonan, 2011). For the management of pink diseases, prune infected branches and paste cut ends with bordeaux paste or copper fungicides (copper oxychloride, 0.2%) (Pathak, 1980; Patil, 2013).

Leaf spot (*Phyllosticta artocarpina, Colletotrichum gloeosporioides*):

It is most common disease of jackfruit. Dark brown spots which subsequently turn into greyish-brown to brick red are commonly observed on leaves (Tanyonan, 2011). Such leaves defoliate prematurely. Tender shoots and petioles of leaves are also infected by fungus. Spraying of Carbendazim (0.1%) or Mancozeb (0.3%), methyl thiephonate (0.1%) or Chlorothalanil (0.2%) is recommended to control (Rawal and Saxena, 1997; Patil, 2013).

Die-back (*Lasiodiplodia theobromae/Colletotrichum gloeosporioides/ Fusarium moniliferme* Sheldon)

It is a highly destructive disease of jackfruit. Drying of young twigs from tip downwards and browning and rolling of leaves are common symptoms. Such branches are infested by shoot-borers. The infected twigs show internal discolouration when split open. In early stages, epidermal and sub-epidermal cells of twigs are often slightly shrivelled. Pruning of infected twigs followed by spraying of Carbendazim (0.1%) or Topsin M (0.1%) or Chlorothalonil (0.2%) is recommended (Rawal and Saxena, 1997; Patil, 2013).

Blossom rot or fruit rot (*Rhizopus artocarpi)*

Blossom rot, or fruit rot attacks the male spikes and young fruits, resulting in premature shedding of tender fruits. In Bangladesh, fruit rot is reported to be the major disease. Ghosh (1994) reported that 18.9% female spikes are dropped due to infection of *Rhizopus*, but it can be treated with the application of folicur or tilt. Fruit rot can be controlled by: (i) pruning trees to encourage good ventilation and to reduce relative humidity in the canopy; (ii) removing and destroing diseased fruits from trees and the ground; (iii) sanitation and preventing waterlogged condition; (iv) controling weeds around young trees; (v) keeping ripe fruits away from contact with the soil or decaying organic material; (vi) avoiding wounding the fruit; (vii) washing fruits after harvesting in clean water and dry thoroughly before packing or transporting; (viii) avoiding storing fruit after harvesting in hot, poorly ventilated containers and; (ix) spraying of Bordeaux mixture (0.5%) or copper oxychloride (0.2%). Though there are reports of dry rot (*Phellinus* sp.), gray blight (*Pestalotia* sp.), charcoal rot (*Ustilina zonata*), anthracnose (*Colletotrichum gloeosporioides*), rust (*Uredo artocarpi*) and root disease (*Fomes durissimus*) (Azad, 2000), these are not of serious nature.

Fruit yield

To achieve the maximum productivity, trees should be 12 years or even older. The fruit yield varies depending on variety, but most varieties yield 100-200 fruits/tree depending upon proper cultural management practices (Amma and Kumaran, 2011). The trees attain their peak bearing stage in 15-16 years after planting. At this stage, normally a tree bears up to 250 fruits annually with annual fluctuation in yield. The weight of fruits also varies depending on the type. On an average, 40-50 tonnes of fruits/ha could be obtained. The fruit yield varies depending on the variety, but most varieties yield 100-200 fruits/tree depending upon proper cultural management practices (Valvi *et al.*, 2011). Fruit bearing is delayed and fruit quality is affected if trees are grown at a higher elevation (Medagoda and Tennkoon, 2001). The flowering to fruit maturity takes about 4-5 months depending on variety and growing conditions. The small-sized fruits mature faster than big ones.

Old unproductive jackfruit tree

Old senile orchard (jackfruit)

(A) New shoots on old trunk (A & B) (B)

(A) New shoots on pruned tips (A & B) (B)

Thinning of shoots

Canopy development

Productive branches on trunk

Branches at pruned tips

Thinning of dense branches

Three branched canopy framework

(A) Flowering, fruiting on trunk and branches (A & B) (B)

Heavy fruiting on rejuvenated tree

Chapter 14

Litchi

Litchi (*Litchi chinensis* Sonn.), a member of the Sapindaceae family, originated in China. An arillate fruit, it is widely cultivated in subtropical regions of the world, providing livelihood security to millions of people in south-east Asia (GalanSauco and Menini, 1989; Menzel and Waite, 2005). Litchi is important commercial fruit crop in India. It is popularly known as "queen" of fruits due to its attractive colour, taste, flavour and excellent quality. Nutritionally, it is rich in sugar content with appreciable content of protein, minerals, fibres, vitamins and calorific value. Litchi is consumed mainly as fresh and little as processed form. The juicy aril is an excellent thirst quencher. It is highly perishable in nature. The short span of fruit availability coupled with poor shelf-life, limits duration of its availability in the domestic as well as international market. Till date, it is meagrely exploited for potential production and due R & D support is still lacking.

PRESENT STATUS

At present annual fruit production is around 585,300 metric tonnes from an area of 84,170 ha and the national productivity is 6.95 tonnes/ha. In India, main production region is concentrated in Bihar, West Bengal, Uttar Pradesh, Uttarakhand, Assam, Jharkhand, Tripura, Punjab, Haryana and Himachal Pradesh. The area under this crop has undergone substantial expansion in the past 64 years with plantation increasing from 9,400 ha in 1949 to 78,000 ha, in 2013 (NHB, 2013). Till date, the present level of production is unable to cope up with the ever-increasing demand. Its fruits are available from early-May to July in different areas.

PRODUCTION CONSTRAINTS

Despite the fact, that litchi is one of the finest fruits and has a growing demand in national and international markets, productivity continues to be low and gap exists between potential and existing yield. The probable reasons for low yield are lack of suitable, superior, high yielding cultivars/varieties, traditional production system and poor technological support. The lack of suitable cultivars and short harvesting season have also inhibited its growth. There is slow pace of area expansion due to shortage of genuine planting material vis-a-vis long juvenile period of litchi plants.The lack of canopy management gives rise to luxuriant vegetative growth, which results in poor yield and causes problems in harvesting (Goren and Gazit, 1992; Huang *et al.*, 1992). Lower level adoption of recommended package of practices, particularly for water and nutrient requirement resulted in poor fruit yield. The yield loss due to physiological disorders and incidence of insects and pests coupled with poor post-harvest management are also major constraints. Among various factors affecting quantum of fruit production and the low productivity is primarily due to the existence of a large chunk of old senile, unproductive and diseased orchards/ plantations. In India, it is about 25-30 % (Kumar, 2007).

Furthermore, lack of region-specific recommended new agro-techniques are also the cause of low productivity, rendering it difficult to compete globally and ensure profitability to growers. Above all, majority of litchi growers have limited access to improved / best cultivation and management practices, which further constraint in litchi fruit production. Proper technological intervention after understanding local production systems and the resources available to farmers within the systems is essential that maximize returns to inputs, reduce seasonality and promote availability / productivity of quality fruits.

REJUVENATION TECHNIQUE

Rejuvenation, as name implies is restoration of youthful vigour. It is a medium term developmental measure, with modest levels of investment. It helps in restricting the deterioration of the health of the plants, thereby extending their life, and aims at improving the yielding capacity of quality production of the existing plants/trees and situation under minimum period of time as compared to establishment of new orchard which attains the commercial bearing stage after 8-10 years.

Rejuvenation of old, dense and unproductive orchards is a medium-term strategy for better productivity with quality of produce. After rejuvenation of old senile orchard, it can be given new productive life for another 12-15 years and their productivity and quality can be restored.

Litchi tree has strong renewable ability and the break even point for rejuvenation pruning with or without considering crop loss is two years, whereas establishment of new orchard take more than 8-10 years to come into profitable bearing stage. So, it is indeed worth to rejuvenate the old senile orchards.

Rejuvenation of old orchard is very vital under orchard management, which needs careful, visionary and scientific approach and it should be followed in skillful manner. In this case, scientists of National Research Centre on Litchi, Muzaffarpur, may be consulted for proper guidance and help, as the technology of rejuvenation developed have successfully come up in farmers field.

Identification/Selection of Old Orchards

In general, litchi plant starts bearing after 5-6 years of planting and attains commercial bearing stage at least after 10 years and commercially viable life (period) lies between 11 and 30 years. At this commercial bearing stage, plants need skillful management by proper pruning and training for stature and canopy management apart from proper nutrition, irrigation and disease and pest management. The orchards after attaining the age of 30-40 years and spaced at 10m x 10m turn dense, compact top canopy cover with bare (devoid of leaves), long-high braches at the bottom and bearing fruits only on the high-tops. They become uneconomical to maintain above 30-40 years of age due to bearing of poor quality fruit at high canopy, pose problem in harvesting and the obtained yield loss due to physiological disorders, incidence of pests and diseases, lack of resources and facilities and develop tendencies of alternate/biennial bearing. It also poses problems in proper management. In some old orchards, the yield is high but fruits fetch very less price due to inferior quality, particularly with respect to size.

Under such condition, it is rather difficult to stipulate the economic longevity of litchi. On the other side, with rising cost of management, it is not economical to maintain these old senile orchards of above 40 years of age. Such orchards need to be rejuvenated for further higher productivity of quality produce. But it is important to look for the health, location and value of trees, which should be determined before going for rejuvenation.

Training to orchard growers

Considering the various pros and cons of rejuvenation, it is necessary to undertake this programme after giving proper training for acquiring knowledge to the orchardist to develop the expert system and apply the methodology in scientific manner. As several questions should be answered before an attempt is made to rejuvenate trees in the old orchards. The important ones are : Is the tree worth saving for fruit production? The orchardist should have equipped themselves with technological development and ideas that litchi tree has strong renewable ability and the breakeven point for rejuvenation pruning with or without considering crop loss or income gap is two years, where as establishment of new orchards take 8-10 years to come into commercial bearing. The harvesting of enhanced quality production like young commercial bearing orchards can further be enjoyed for at least another 12-15 years. Proper training is required for increasing awareness, which may also include aftercare treatments like: (i) efficient management of inputs like nutrients, water; (ii) proper intercultural operations by taking even suitable intercrops; (iii) thinning operations and canopy management and; (iv) management of pest-diseases. This type of pruning is normally undertaken on old trees which are still healthy with the aim of reinvigorating the tree to improve cropping potential.

Reiterative Pruning

In rejuvenation, heavy reiterative pruning of litchi tree is done at the height of 2.0-2.5 m depending upon the girth and type of main trunk. About 3-5 main branches with outward growth from the base are marked for pruning at this required height, with the plan of developing umbrella like or semi-circular framework of tree canopy. Pruning can be done either with manual saw or power operated saw. Care should be taken to avoid bark splitting or debarking at the cut end due to falling of heavy branches at the time of pruning. It is always advised to go for reiterative pruning for rejuvenation in phased manner starting from the top light pruning following top branch cutting then only the final cut at the marked point. Aftercare is very important for successful rejuvenation. To avoid any external infection at the cut portion, it should be pasted with Bordeaux mixture or copper-oxy-chloride (Blitox) immediately after pruning,

Time of Pruning

By adopting and following all the required scientific techniques the best time to go for reiterative pruning for rejuvenation is August –September i.e. mainly after the rainy season. Rejuvenation is labour intensive operation and it is extremely necessary to complete the operations during August-September to achieve the desired results.

Application of Manures and Fertilizers

After the reiterative pruning, trees should be fed with optimum dose of fertilizers (Menzel, 1987), followed by irrigation near the root zone. The manures and fertilizers should be applied through ring method. The dose of an adult bearing tree is 75-80 kg well rotten FYM, 2 kg neem/castor cake, 1kg Urea, 1.5 kg single super phosphate and 500g muriate of potash should be applied preferably in two split doses: one during August-September and another in February-March. The micronutrients like zinc ($ZnSO_4$ @ 2-3g/litre) and boron (Borax @ 2g/litre) may be applied through foliar application one year after the reiterative pruning (Kumar, 2008, 2012, Menzel and Simpson, 1990).

Old senile unproductive litchi tree

Pruning of tree branches

Pruned tree

Emergence of shoots

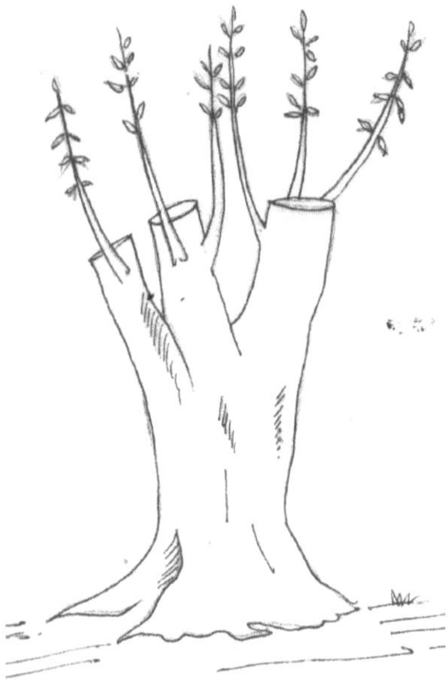

Thinning of shoots

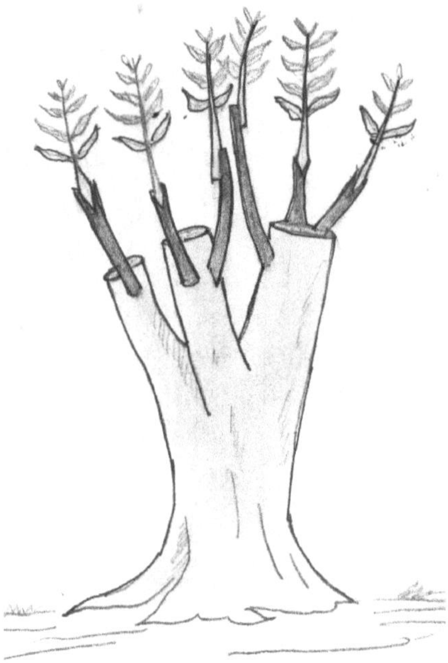

Top grafting

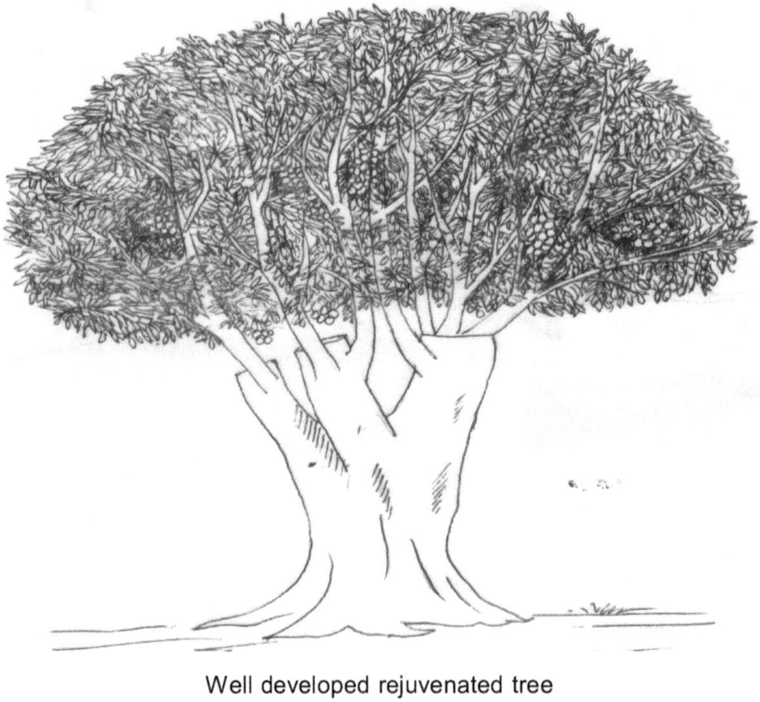

Well developed rejuvenated tree

Thinning and Canopy Development

The pruned trunk/branches in August, starts putting forth vegetative sprouts just after 25-30 days variety. Shahi and 40-45 days in China. The enormous number of shoots come out and cover the entire body of the pruned plant. At this juncture, carefully thinning operation should be done with the ultimate aim of developing solid, semi-circular or open umbrella type canopy. Now, cut away suckers around the bottom of trunk and thin out the excessive branches as well, keeping only 3-4 numbers nearly top-side portion on each pruned limbs of trunk projecting all the directions and further training for acquiring desired shape. In India, top grafting is not being practiced but rejuvenation by top grafting is being encouraged in China and South Africa. The air and light must have access to all the parts of the tree. The horizontally developing limbs should be cut back to the point where they do not hang under the weight of fruits into row, on to the ground of on to lower limbs.

Irrigation

Rejuvenated trees require immediate irrigation in case of dry spell. Irrigation is must after application of manures and fertilizers. Irrigation at regular interval of 8-10 days during hot summer season and 15-16 days in winter season helps in fast vegetative growth and good canopy development. Irrigation applied to the intercrops also gives the added advantage to the rejuvenated trees, apart from maintaining/improving soil physical condition (soil health) and complete control over weed growth. Mulching during April-May and September-October have been beneficial in conserving soil moisture beneath the trees, reducing the frequency of irrigation and enhancing growth rate of vegetative flush.

Intercropping

Just after the completion of reiterative pruning for rejuvenation, the open interspaces between the plants/trees is created in the orchard like newly-planted orchards. Intercrops in summer season like blackgram, greengram, maize, cucurbitaceous vegetables and fodder crops and cowpea, frenchbean, pea, potato and mustard in *rabi* season are most suitable. Flowering plants and annual fruit crops like papaya and banana have also give good income for 2-3 years. Apart from the significant income obtained by intercrops in rejuvenated orchards, improvement in the soil tilth (health), almost complete control of weed population/ growth as well as less incidence of pests and diseases are added advantage.

Management of Physiological Disorders

Litchi cultivation is threatened by many physiological disorders. It is important to understand their causes and symptoms properly to manage them. The incidence and severity of these physiological disorders vary with locality, season, cultivar and orchard management practices (Kumar, 2008).

Fruit cracking

The fruit skin-cracking (splitting) is most important disorder in litchi. It occurs in almost all important litchi growing countries in the world. It causes 5-70% losses. It is a associated with hot, dry weather, drought and low calcium concentrations. It is observed that early ripening cultivars observing poor management practices are more susceptible to it. Deficiencies of nutrients can affect fruit development causing fruit cracking. Role of calcium is also found associated with fruit cracking. But litchi-growing areas in Bihar soils are rich in calcium content compared to other states. The other nutrient element found to be associated with fruit cracking is boron (deficient soils). The relationship between fruit cracking and endogenous growth substances have been also established. The higher concentrations of abscisic acid (ABA) and lower level of gibberelins in fruit pericarp, seed and aril of cracked fruits. Insects, hailstorms, and the sunshine can damage the skin during cell expansion and induce cracking towards harvesting.

In India most of the cultivars are found susceptible and it. The susceptibility depends on environmental, hormonal, varietal, nutritional, soil moisture, disease and pests, and mechanical injuries. In general, fruit skin cracking often occurs when trees are droughted soon after fruit setting. If the drought is severe enough, fruit development is affected, particularly the development of fruit skin. Cell division is reduced and the fruit skin becomes inelastic, and often splits when aril grows rapidly before harvesting. This can occur after irrigation or heavy rain, or just an increase in relative humidity.

The cracking injury in litchi may vary from short cracks, generally skin deep to larger ruptures extending to most of the length of the fruit. As the underlying flesh (aril) gets exposed to dry air, it withers rapidly. The cracks on the skin may be longitudinal or horizontal. The cracked fruits deteriorate rapidly and thereafter suffer secondary infestation by disease. They become non-marketable and ultimately cause great economic loss to the growers.

Management : Application of calcium (@2ml/litre liquid formulations) and gibberellins (@20 ppm), which reduce the activity of cellulose and thereby reduced cracking (Chandel and Kumar, 1995; Bhat et al., 1997, Sinha et al., 1999; Peng et al, 2001). It has been also found that application of 2,4-D and NAA at concentrations @ 20 ppm reduced cracking (Chandel and Sharma, 1992).

Spraying of Borax or Boric acid @ 2g/litre at the initial stage of aril development confirming enough soil moisture in the root zone checked fruit cracking significantly (Kumar & Kumar, 2008).

Maintenance of moisture and humidity properly are needed at the time of fruit maturity. Irrigation at 30-40% depletion of available soil moisture is quite helpful in reducing cracking of fruits. The soil moisture conservation in the root zone for prolonged period can be achieved by providing mulching beneath the canopy spread area. There is also need to develop cracking resistant cultivars as only one cultivar. Swarn Rupa developed is tolerant to this disorder. Early cultivars or those particular cultivars which have relatively thin skin few tubercles per unit area and rounded to flat in shape are less prone to cracking can be selected for commercial plantation.

Planting wind break around the orchard provides protection from desiccating hot winds. Suitable wind break should be planted around the orchard boundary and it should be at a right angle to the direction of prevailing wind. A row of tall growing trees, such as seedling mango and jamun are suitable wind breaks

Sunburn

Sunburn (lesion browning pericarp necrosis) in litchi group is a serious handicaps in litchi producing countries like India, South Africa, Australia and Tahiland (Kanwar *et al.*,1972b; Joubert, 1986; Sanyal *et al.*, 1990; Menzel *et al.*, 2002; S. Rasnanda, 2002). The damage caused due to sunburning occurs upto 0.9-19.13 % in different varieties.

Climatic factors are determinants for incidence and severity of sunburn disorder. Besides, varieties, hormonal, nutritional and soil moisture factors have also been found associated with this sunburn (Chadha and Rajpoot, 1969; Joubert, 1986; Sharma and Ray, 1987; Sanyal *et al.*, 1990; Menzel *et al.*, 2002).

It is physiologically related with PPO (Poly-Phenol Oxidase) activities and it also varies with cultivars. Sunburn is pronounced in ill managed orchards having sandy or sandy loam soils or light soils receiving/exposed to high temperature (>40°C) and very less RH (<50%). It is a type of direct thermal injury. At higher temperature the tissue or organ in question coming in contact/exposure gets sunburnt/sunscalded. Sunburn problem is also experienced more in early ripening cultivars. Fruits on shaded branches suffer less damage than those in the sun (more exposed to sun). Lower translocation of calcium in the pericarp region also found to favour sunburn disorder. The fruits in orchards having scanty irrigation during fruit development stage receiving rain drops on the fruits favour sun burning.

The light brown blotches appear on fruit skin facing direct sun rays. In severe cases more than half of the surface area becomes discoloured, blotchy light brown. The blotches become intense in few days and the blotchy area dried up blocking aril growth and finally destroying the market value of fruits. The symptoms appear more on the south-west side than the north-east side as on the latter side fruits remain almost in shade except during early hours of the day, which is not harmful. Under similar climatic conditions the bearing of the fruit may suffer with this disorder due to the position of the fruit on the tree.

Management : Irrigation at regular interval during the fruit growth and ripening stage may keep the trees and fruits cool to escape sunburn. Irrigation through sprinkler system in morning and evening hours increased humidity, cool the orchard atmosphere, decreasing the incidence of sunburn. In light and sandy soils only light irrigation with increased frequency (4-5 days interval) reduces the chances of sunburn in farmers' fields.

The trees fed with sufficient quantity of organic manures, particularly compost, FYM, cakes, green manure, vermicompost, and applied irrigation at regular intervals during fruit development and ripening stage showed almost no symptoms of sunburn. Planting windbreak around the orchard reduces the sunburning problems. Therefore, suitable windbreak should be planted around the orchard boundary.

Flower and fruit drop

Sometimes even after profuse flowering and fruit setting, poor yield is observed because litchi trees suffer a heavy flower and fruit drop between flowering and fruit maturity. Only a small proportion (2-18%) is carried to maturity in different cultivars. The quantum of flower and fruit drop varies with variety, season and tree age.

The main cause of flower and fruit drop is imbalance of nutrients and hormones needed for embryo and fruit development, lack of pollination/fertilization and environmental conditions in general and in certain cases due to fruit borer and heavy mite attack. The younger tree suffers less drop than older trees. The drop of fruit is thought to be due to failure of fertilization, embryo abortion, nutrition and hormonal imbalance and external factors like high temperature, low humidity and strong westerly winds (Chadha and Rajput, 1969; Menzel, 1984, Sharma and Ray, 1987). Maximum fruit drop during the first fortnight after fruit setting and at harvesting the retention varied between 3-39.6% among the cultivars. It is postulated that the high incidence of fruit abscission could be a physiological rather than genetic problem. The auxin inhibit the action of the hydrolytic enzymes polygalacturonase and cellulose, which are responsible for degradation of the cell wall and middle lamella in the abscission zone (Goren, 1993). Though, specific relationship between auxin production in developing fruit and abscission in litchi has not been clearly investigated but many studies have been conducted to examine the possibility of reducing abscission in litchi by application of synthetic auxins.

In litchi, heavy fruit drop is observed at different stages due to the formation of abscission layer. Although fruit drop continue up to maturity, most of flowers and fruit drop occur during the first month after the pollination. The male and hermaphrodite flowers (behaving as male) only provide pollen and fall down. The female flower only borne fruits after pollination and fertilization processes. The initial fruit setting in litchi is very high but a very small proportion finally mature. The premature fruit drop commences soon after fruit setting and continues till fruit maturity, with most fruit abscising in the first 2-4 weeks.

Management : Fruit drop control differs in different states with varieties. High fruit retention in litchi can be ensured by controlling winter shooting, flowering period and flower quality. Timely treatment of plants for strong and healthy shoot production delays flowering phase and increases the female flowers and finally the fruit setting rate. Honey bees are the main pollinating agent in litchi. Increasing bee population in orchards ensures better pollination and fertilization which increases the fruit setting and fruit retention. Do not spray at the blooming stage in the orchard.

Spraying of $ZnSO_4$ @ 0.5% 30 days before panicle emergence induces healthy inflorescence leading to more fruit setting. Foliar application of 0.4 % boron, 2-3 times during the period of fruit growth and development enhance fruit retention, minimizes cracking, improves fruit colour, sweetness and enhances maturity. Two foliar application of plant bio-regulator (NAA @ 20 ppm; planofix @ 2ml/5 litre water) at an interval of 15 days from peanut sized fruits is useful in minimizing the fruit drop effectively. The IAA may be substituted for NAA. Treatment with growth regulators like NAA at 20-30 ppm, GA_3 at 20-25 ppm, 2, 4-D at 10-20 ppm are effective in minimizing fruit drop when sprayed on panicles, before the flower opening. The fruitlet drop can be reduced by spraying of $ZnSO_4$ and synthetic auxins like 3,5,6-trichloro-2-pyridyl-oxyacetic acid (3,5,6-TPA) on fruits increasing yield.

Irregular bearing

Low and irregular bearing is a serious problem in almost all litchi growing areas of the country and abroad (Mustard *et al.*, 1956). The capacity of litchi shoots to bear is varietal in nature. While, some bearing terminals of current year are more productive (70-95 %) in the next year. The floriferous condition of shoots is determined largely by presence or absence of crop in previous year, rather than by how many times a shoot grow vegetatively before its differentiation into a floriferous condition. It is also been observed that adoption of faulty management practices lead to irregular bearing, even in regular bearing, even in regular bearing cultivars. Faulty management practices include late application of manures and fertilizers (particularly nitrogenous fertilizers), high frequency and heavy irrigation just before the panicle emergence and flowering period. In cultivar, China, which is generally known as irregular bearer, even bear fruits every year by adopting the improved management practices timely and accurately. Irrigation, nutrition and environmental factors are also responsible for the cause of irregular bearing in litchi. Type of soil coupled with management practices are associated with irregular bearing or alternate bearing. Late harvesting and severe pruning and training operations may lead to non-bearing in that particular year of operation.

Usually, heavy crop is followed by a light crop in some varieties, while in some varieties production is quite consistent every year. In some cases, at the time of bearing panicle/inflorescence, trees start giving vegetative flush and that particular year becomes off year, while in some cases only part of the tree bear long panicle with leafy growth at the base, resulting only few flowering terminals and fruit setting is also very poor. It is also observed that inspite of heavy bearing of flowers and enough fruit setting, harvest of mature fruits is very poor due to heavy flower and fruit drop at various phases during flowering to fruit maturity.

Management : For sustaining commercial cultivation of litchi in India and elsewhere, varieties chosen for planting must be regular in bearing, apart from other desirable characters like heavy yield, intense red colour of the skin, high fruit quality, long shelf life at ambient conditions and good transportation and storage characteristics. In India litchi industries depend largely on Shahi, Dehradun, China (Calcuttia), Purbi, Bombai, Rose Scented varieties, while Dehradun is a regular bearing, but China, Seedless, Late Bedana and Rose Scented are irregular/shybearers.

On the basis of age of the trees, the commercial orchards should be properly fed with balance nutrition immediately after fruit harvest and irrigation at right time through right method, particularly at aril development stage. The plant protection measures to control major pests and diseases should also be followed.

Litchi tree requires proper pruning and training at the time of harvesting and just after harvesting, while harvesting of the fruit bunches, the branches having 8-10 inches length should be plucked carefully, to give rise strong and healthy flush to bear fruits in next year. The trees should also be trained to give canopy shape in a semicircular manner or an open umbrella shape. Late pruning and training should be avoided.

Retarded / underdeveloped fruits

This disorder is not well documented but somewhat causing considerable loss due to poor quality fruits (Kumar and Kumar, 2007). More prevalent in poor managed orchards coupled with poor pollination decide the severity of this disorder.

Lack of proper fertilization by less visits of pollinators may lead to improper fruit set and retarded fruit growth. Insect-pests attack and physical jerk also cause retarded or underdeveloped fruits. Unavailability of moisture during development stage of fruit is also one of the cause for retarded/underdeveloped fruit.

The fruit size is smaller than the normal ones. The juice content or aril development is also very poor. This type of fruits remain in glossy pale green colour for a longer period and then only dull reddish colour appears and just after the problem of prematurely fruits drop also starts. During the fruit growth of litchi, seeds develop initially at a high rate, followed by membranous mesocarp and aril (or arillus), which grow very fast towards the later stage of fruit development. During these phases, sometimes it happens that there is no aril development or aril development is incomplete with portion of testa uncovered also give rise to the development of retarded/underdeveloped fruits.

Management : Putting enough number of honey bee colonies during flowering to fruit setting stage increased fruit setting with better fertilization leading to normal and healthy fruit growth. Spraying plain water in early morning hours of the day during the advanced stage of growth and development have been found highly effective in better growth of fruit quality. Arrangements of sprinklers system of irrigation also found to be effective in reducing this disorder.

Black tip

The occurrence of black tip is not wide spread in litchi (as in mango) but it has been found infesting and damaging the quality harvesting of quality fruits.

In Indian condition, this symptom has been observed in plantation mainly in vicinity of urban areas (Kumar and Singh, 2008) and no other reports about its occurrence in litchi. It is supposed that it is caused due to deleterious effect of smoke fumes which contain sulphur dioxide, acetylene and carbon dioxide as found in case of mango. The symptoms of black tip becomes apparent when the fruit attain certain size and remain in developmental phase i.e. from March-end to mid-May. The first symptom appear as the development of a small etiolated area at the distal end of the fruit against the normal green colour of the fruit pericarp, which gradually spreads, turns nearly black and covers the distal end completely. The infested fruit show discolouration and overall slow pace development resulting very poor development of aril, seed, size and colour.

Management : It may be managed by applying proper nutrition and irrigation (particularly during fruit growth and aril development stage). Spraying with micronutrients like zinc ($ZnSO_4$ @ 0.2%) at one month before flower panicle initiation and boron (borax @ 0.2 %) during fruit growth stage can escape the infestation.

Heat injury

It is more severe in young plantations of smallholder farmers' fields as compared with the commercial orchards. The severity of each varied between different litchi cultivars and areas with different agro-ecologies. Higher temperature, soil type, high rate of respiration in comparison to that of photosynthesis causing stunted growth, imbalance in moisture content in root zone etc., disturb and destroys hormonal mechanism and synthesis of essential metabolites and their translocation. The morphogenic changes, apical dominance, thermo-periodic induction etc. are adversely affected by higher temperature or wide temperature fluctuations. All together these adverse prevailing situation force the young growing litchi plants or even the bearing tree to a permanent wilting situation or may lead to death.

Management : Application /integration of proportionate amounts of biofertilizers, organic manures and reduced dose of chemical fertilizers, all IPNM system and timely application. Irrigation during fruit growth and development stage and maintenance of enough moisture content in rhizosphere, mainly in poor fertility status of soils or sandy to sandy loam soils are of vital importance.

Salt injury

Salinity/alkalinity problem is another important factor causing injury to litchi plants. In problematic soils, when excess of boron /$CaCO_3$ are present, it causes deficiency of Zn, Mg, Cu etc., these salts are present in excessive quantities in soils, tree growth and performance are negatively affected. The leaves are scorched and loose their natural colour and turn to bronze colour. Tip burning is also seen in some plants due to salt injury.

Management : Raising *dhiancha* (Sesbpania) as a green manure crop with onset of monsoon in the interspaces and its incorporation in soil during September have been found effective. Farmyard manure and compost should be applied adequately every year. Application of gypsum and scheduling of irrigation water reduces salt effect. Frequent irrigation and mulching of soil are important during vegetative growth and fruit development phase.

Management of Pests

The new re-growth and development of canopy for initial bearing phase in rejuvenated orchards have the same effect and infestation of pests as in existing orchards. As this crop is attacked by number of insect pests, posing threats to lucrative harvest of fruits and therefore its effective and eco-friendly management to control is essential.

Erineum mite (*Aceria litchii* Keifer)

The erinose or leaf-curl mite (*Aceria litchii* Keifer) is a major litchi pest found damaging litchi in almost all the litchi growing countries. Both nymphs and adults stage infest leaves, inflorescence and young developing fruits. causing hairy, blister-like galls on the upper side of the leaves, thickening, wrinkling and distorting them, and brown, felt-like wool on the underside. The infestation generally begins from the lower portion of the trees and gradually extends upwards. The attacked leaf buds fail to bear flower or fruit. In India, the attack of this mite is much severe on Shahi, Bombai, and China varieties, while cultivars like Kasba, Trikolia and Longia show tolerance to the pest. Plant propagation from mite infested mother plants is the major cause of rapid spread and the other viable cause of spread are touching of the infested tree with the trees in vicinity, transfer by human activity, other insects movement (agents like honey bees during pollination), and also by way of wind velocity and movement (Prasad and Singh, 1981; Wen, *et al.*, 1991; Waite, 1999).

Management : The control measures must be preventive, for once the mite is established, it is almost impossible to eradicate. It is recommended that plant propagation for making layers be prepared only from non-infested plants. The leaves should be checked regularly for symptoms over summer and autumn. Branches infested with the mite should be cut off and burnt. In September-October, trees must be treated just prior to vegetative flushing with 0.05% Dimethoate either alone or in combination with 0.12% Dicofol. After harvesting in June, infested branches must be removed during pruning. Sprays of wettable sulphur have been recommended for the control of erinose mite (Wen *et al.*, 1991; Waite, 1992b). Spraying of insecticides like Kelthane (1.25ml/litre) or Dimethoate (1.25ml/litre) or Phosphomidon/Nuvacron (1.25ml/litre) is recommended to control mites. Numerous species of predatory mite have been recorded with *A. litchii* Keifer in all important litchi growing countries. In India, the species of *Phytoseius* sp., *Typhlodromus* sp., (Lal and Rahman, 1975), *Amblyseius* sps. (six species), *Typhlodromus* sp., (Somchoudhary *et al*, 1987; Thakur and Sharma, 1989) are found associated and preying *A. litchii* and used as biological control.

Litchi fruit borer (*Conopomorpha cramerella* Snellen)

The litchi fruit borer or cacao moth also known as litchi stem-end borer or litchi seed borer, is the major pest in India, Thailand and China.

The newly hatched larva feeds on the fruit skin and then tunnels towards the seed. In immature fruit, the young larva bores directly into the seed, which is completely eaten. In early stage of infestation, newly hatched larvae start boring into the fruits and feed on its pulp. The infested fruits do not attain normal size and can be identified by the formation of black spot near pedicel. Larvae do not enter much deeper into the pulp but feed below the calyx about 15mm deep and in case of severe infestation, the loss caused due to heavy fruit drop.

Management : Infested fruit should be picked and destroyed, at infestation levels of 1-2%. Bagging of fruit panicles after fruit set by nylon mesh bags can exclude moths or alternatively, the panicles can be covered with paper bags. The bags also improve fruit colour and overall quality. Neem based products may be applied at the time of new shoot emergence to avoid heavy population of the pest. Spraying 0.05% Fenitrothion or Dicholorvos or Carbaryl 2g/litre may be done. The insecticides like Cypermethrin, Deltamethrin, Carbofuran or Fenthion during early fruit setting is recommended to prevent damage later in the season.

In India, the egg parasitoids, *Trichogrammatoidea fulva* Nagaraja, (parasitised egg cards, i.e. Trico-cards hung in trees during February-March and April-May in orchards under integrated pest management, used in the field trials has given encouraging results) was considered to be the most important of a number of natural enemies of *C. ombrodelta,* parasitizing upto 68% of eggs (Annonymous, 1976). The trial conducted at NRC on Litchi, Muzaffarpur (Bihar) has recommended the use of *Trichogramma chilonis* @50000 eggs/ha and use of pheromone trap, however, this pheromone is not giving consistent results in all the situations. It is highly recommended that the fallen fruit should be removed to reduce the build-up of moths.

Bark-eating caterpillar (*Indarbela quadrinotata; Indarbela tetraonis*)

It is a serious polyphagous pest of litchi in India and China (Zhang, 1997), causing significant damage to the litchi trees. Litchi is damaged to a considerable extent by the bark-eating caterpillars, which attack trees of all ages, particularly the older ones, lowering their vitality. They bore into trunk, main stems and thick branches of litchi trees. The old, shady and neglected orchards are more prone to attack by this pest. When severely infested, the entire branch or tree may die. The larva is the damaging stage of the pest and this beetle has one generation per year, with adults emerging from June to August. Tunneling by the larvae may kill branches, but rarely whole trees. The caterpillars feed on the bark of the tree during night. They bore inside the trunk or main stems and eat through the bark into the wood, chew wood particles and weave cotton like webs, containing their excreta (frass), which hang on the trunk near the crotches of the main branches. The affected portion is covered with large silken webs. Ring-barking of twigs by ovipositing adults causes the shoot tips to die and snap off. In case of severe infestation, sap movement is affected due to which the growth of the tree is restricted and fails to flower. In severe infestation tree decline takes place.

Management : The infestation can be can be located from the appearance of their frass, which is packed into the ends of tunnels. A small metal wire or spoke is inserted into the hole to fish-out or kill the larva/caterpillar. The holes are then plugged with cotton wool soaked in fumigants or chloroform or formalin or petrol or kerosene oil and then the opening of tunnel is sealed with mud. Before going for control measures the covering of fresh frass and webbing must be cleared by brooming the affected surface to find the hole (infestation point). The operation is repeated at least two-three times (September-October and January-February). In China too, Dichlorvos is injected and the tunnels sealed with clay (Zhang, 1997). Putting emulsion of DDVP (0.05%) in each hole and plug them with mud. Mix Chloropyrphos 2 ml/litre of water and apply the bark eating caterpillar infested area with a brush at 15 days interval.

Leaf roller (*Pletypeplus aprobola* Meyer)

Leaf roller has been recorded as minor pest in India. This pest attacks leaves, mostly vegetative flushing and flowers. In Indian condition, the incidence of damage is found from July to March when new leaves/flush are available. This insect rolls leaves and feeds on them during the vegetative flushing. It attacks flowers during flowering period . As a result of larval injuries, the infested twigs distort and wither. Litchi trees whose foliage is attacked by the larvae, very poor flowering or in case of younger tree, no flowering is seen in the season. Thus, the crop yield gets reduced considerably.

Management : In India, rolled leaves that contain larvae are removed manually during infestation. In severe infestations the spraying of insecticides Phosphamidon or Fenitrothion (0.03%) is applied. The spraying with Carbaryl @ 2g /litre have been found effective to minimize damage to young trees or at critical periods of leaf growth in older trees, when applied at 20% of leaf flushes infestation.

Leaf miner (*Acrocercops cramerella* Snell/*Conopomorpha litchiella* Bradley)

It is an important litchi-pests growing in Asia-pacific region It cause damage by mining the leaves mainly mid rib and veins. It is creamy white and the moths phase is attracted to newly emerged leaf flushes. The female moth of this insect lay light yellow eggs on new shoots. The creamy white larvae bore into the shoots, but they may also mine the leaf blade, although mature larvae prefer to feed on the mid rib and veins of young leaves, they may also bore into fruit during the fruiting season. The leaf-miner is attracted to leaf flushes, especially during the rainy season and onwards, i.e. from July to September or even up to October. New tender twigs and leaves turn brown-black and wilt.

Management : Orchard should be kept clean. Bearing trees should be inspected for the presence of leaf miner during major flushes. At least 2-3 sprays of Cypermethrin (0.01%) or Monocrotophos (0.02%) must be applied to very young leaves in between July to October to manage the pest.

Stink bug (*Tessaratoma papillosa*)

The bugs belonging to Tessaritomidae as emerging pest, attack litchi throughout Asia and Australia (Butani, 1977; Singh and Sharma, 1961; Nair, 1995).

It is a large, active insect with bright-red markings which attacks the fruit. In spring, the females are attracted to trees with new flowers and shoots. Both adults and nymph feed on tender plant parts such as growing buds, leaf petioles, fruit stalks and tender branches, and cause damage by sucking the sap from young twigs, which often die and cause a high rate of fruit-shedding. Liu and Lai (1998) claimed that up to 30 percent of fruit in commercial orchards are damaged despite chemical applications.

Management : This pest is combated by shaking the trees in winter, collecting the bugs and dropping them into kerosene. Without such efforts, it works havoc and damage cause a significant loss. In chemical control, timing of sprays is critical because the bugs vary in their susceptibility to insecticides at different times of the year. So, Cypermethrin (0.01%) or Monocrotophos (0.02%) should be applied. Many of these bugs may be controlled with the sprays of either Dimethoate or Fenthion also.

Weevils (*Cratopus angustatus* Bohemann./*Cratopus humeralis* Bohemann)

Weevils are polyphagous and its many species found active during winter in India, attacking litchi trees (Singh, 1971b). Population of the weevil are largest during November-February and smallest during April-July. Heavy infestation results in damage to most of the leaves (even found feeding the tender leaves) and is responsible for low yield in subsequent seasons. The weevils also found feeding on leaves, flower panicles and fruit just after fruit set (Vayssieres, 1997).

Management : Chemical sprays can be made when weevil infestations are noted on the trees. Spraying with 0.04% Diazinon or Dichlorvos or 0.05% Fenitrothion gives an effective control.

Fruit flies (*Ceratites capitata* and *Pterandrus rosa*)

This insect cause damage as pest in India. The female fruit flies lay eggs on the surface of the fruit and the larvae may penetrate weak areas of the skin infesting the flesh or may make minute holes and cracks in the skin and cause internal decay. The fruit flies may lay their eggs through the skin of the fruit, often utilizing cracks and wounds made by other pests. These pests are so detrimental that growers have adopted the practice of enclosing bunches of fruit clusters (bagging). Caribbean fruit fly has become an important pest of quarantine concern *i.e.* at post harvest level also. Plant parts liable to carry the pest in trade/transport include fruits that can house eggs and larvae internally; growing medium accompanying the plants in which pupae can reside and flowers and/or inflorescences, and also the major risk is from the import of fruit containing larvae, either as part of cargo or mail (Baker and Cowley, 1991).

Management : The use of pheromone baited traps for monitoring fruit fly populations around litchi orchards. One of the most effective control techniques against fruit flies in general is to wrap fruit, either in newspaper or a paper/ plastic bag. Spraying of neem based organic insecticides or neem seed kernel extracts as deterrent (Shivendra and Singh, 1998). Parasitoids of *B. cucurbitae* (include *Neoaplectana carpocapsae*; *Opius fletcheri* ; *Pysttalia incisi*) attacking larvae can be used to monitor fruit fly population control. Hot water emersion and vapour heat treatments are also recommended for fruit fly control (Yang *et al.*, 1996). The utilization of pre- and post-harvest management practices, important to reduce direct losses and to increase efficacy of post-harvest quarantine treatments which include mechanical control, cultural control, biological control, and chemical control (Ronald and Jayma, 1991).

Litchi mealy bug (*Drosicha mangiferae* Green)

Distribution : It sporadically attack litchi in India, China, Thailand. Nature of damage : The insect has been reported to damage the litchi plant. The nymphs and adult female are flat, oval and waxy whitish. Males are harmless and females cause damage by attacking on newly emerged shoots, leaves and inflorescences. They suck the cell sap from attacked portion of the trees. The infected part become week, attacked shoots dry up and in case of young plants stunted growth is seen. It may contribute to sooty mould.

Management : The banding of tree trunk with 30 cm wide 400 gauge polythene/ alkathene strip combined with grease on 5cm area on its lower end and plugging the space between the trunk and band effective control the pest. The spraying of Phosphamidon or Dimethoate are recommended for control (Bhutani, 1977).

Thrips (*Dolicothrips idicus* Hood)

This insect attack the foliage and flower in litchi in India and China. All life stages feed on the young shoots causing malformation of the new leaves. Mainly nymphs and adult scratch the epidermis and suck the sap oozing from the wound. The insect population and there infestation found at the apex during August to October. The other strain of thrips also damage the new leafy flush and flowers.

Management : The proper orchard management w.r.t nutrition, irrigation and canopy to encourage uniform flushing. The spraying of phosphamidon or dimethoate are recommended for control (Bhutani, 1977).

Aphid (*Aphis spiraecola*)

Also known as citrus aphid, attack litchi in India (Kanwar, 2002). They cause sporadic damage to litchi plants. It is a polyphagous pest and its culter of adults and nymphs attack flush , foliage and infestation is caused by sucking the cell sap and as a result discolouration of young leaves takes place. The characteristic secretion of honeydew, leads to the development of sooty mould infestation. Leaves covered with black coating material interfere with the photosynthetic activity. Growth and yield potential disturbed. In India, the most favourable period of apex population and infestation is dry weather after rainy season (Bhutani, 1977).

Management : The proper orchard management w.r.t nutrition, irrigation and canopy to encourage uniform flushing.The spraying of phosphamidon, monocrotophos or dimethoate (@ 0.03%) are recommended for control. In case of severe attack the spraying can be repeated at 8-10 days interval (Bhutani, 1977).

Nematodes

Damage caused due to various nematodes have been reported in South Africa (Meline *et al.*, 1971), China (Yin *et al.*, 1994ab; Zhang, 1998) and India (Singh and Khan, 1997). The two major parasitic nematodes *Hemicriconemoides mangiferae* Siddique (called as annular/ring nematode-external parasite) and *Xiphinema brevicolle* Lordello & JaCosta, *X. siamense* nov. and *X. arunachalensis* (known as dagger nematode-system parasite) are causing significant loss to litchi industry in South Africa (Milne, 1982). The attacks of the two other root-knot nematodes, *Meloidogyne javanica* and *Trichodorus sp.* have been recorded also in South Africa but less prevalent. The nematode *Hemicriconemoides fujianesis* sp. Nov which is very close to *Hemicriconemoides mangiferae* Siddique, *Hemicriconemoides litchi*, *Hemicriconemoides birchfielai* reported in China (Zhang, 1998).

Various nematodes have been associated with litchi decline causing mainly dieback in countries like South Africa, China and India. They cause dieback, decline and ultimately death of litchi trees, sometimes devastating orchards. The distribution, root injury and spread of nematode is difficult to detect at early stage of infestation. These attack litchi roots extensively and mainly feeds on root tips, penetrates through root causing the collapse of the cortex and cortical parenchyma. Consequently have a much reduced root system compared to healthy tree, causing plant to stunted growth and finally death. The morphological phenomenon is characterized by ultimately leaf abscission, bare twigs and branches in the canopy and inferior bearing of crop load.

Management : Orchard soil should not be used for litchi plant propagation like marcotting/air layering. Pre-plant soil fumigation with chloropicrin or methyl bromide or chloropropane and the application of post plant nematicides like DBCB (28 or 45 l a.i./ha) or CIBA 14421 (20kg/ha) have shown considerable promise for controlling these nematodes.

Apart from these important and significant damaging pests, there are may non-insect or mite pests damaging the litchi crop but not upto a significant level but in cases they cause considerable damage too, as have been invariably reported from the important litchi growing areas of the country and aborad.

(i) Red ants (*Cecophylla smaragdina* Fabricius)-It attacks litchi and also several other fruit crops in India, Africa and Australia. The ants make nests out of leaves, which they sew together at the stem tips. They also pose great difficulty in fruit picking by painful bites to the pickers and this pest can be controlled by destroying the nest or applying 0.1% BH (Hymexazol).

(ii) Red spider mites *(Paratetranychus hawaiiensis)*-The foliage is sometimes infested with red spider mites.

(iii) Birds, bats and bees damage ripe fruits on the trees (in China, Australia , Florida). To control this sometimes a stilt house is built beside a choice litchi tree for a watchman to keep guard and ward off these predators, or a large net may be thrown over the tree.

(iv) Lac insect (*Kerria lacca*)

(v) In Florida, birds, squirrels, racoons and rats are prime enemies. Birds have been repelled by hanging on the branches thin metallic ribbons which move, gleam and rattle in the wind. Grasshoppers, crickets, and katydids may, at times, feed heavily on the foliage.

Management of Diseases

Litchi plants are affected by numerous diseases but the biology of the diseases are not extensively investigated (Coates *et al.*, 2006). The recent studies showed that the litchi is now subject to relatively few serious diseases, causing considerable loss by lowering grade and quality of fruits and reducing yield. Mainly fungal and algal diseases like e.g. leaf spot, twig blight, flower blight, panicle blight, fruit rots and die back have been reported. No bacterial and viral infections have come to the notice in this crop. Some post harvest rots commonly occur in injured fruits causing major economic losses in marketing and export of fruits. As litchi has great export potential, it is necessary to have knowledge about its diseases and their management strategies.

Leaf spot (*Botryodiplodia theobromae* Pat; *Colletotrichum gloeosporioides* Penz.; *Pestalotia pausiseta* Sacc; *Microdiplodia litchi* Kunze; *Phyllosticta* sp.)

In litchi leaf spot disease is caused by various fungal pathogens. Leaf spot disease is a major problem in most of the litchi growing areas. The symptoms produced by them show several commonness, though the infection is a complex one being caused by either jointly by all the organisms. The anthracnose leaf spot is most common and causing the highest damage and yield loss. The pathogen survives on the leaves throughout the year and develops vigorously after onset of the monsoon. The leaf necrosis type of symptoms caused by this fungus infestation just after the insect injury or mechanical damage. Fast spread of the spores of the pathogens to cause further infections as high temperature and high relative humidity are highly conducive for the development and spread of anthracnose leaf spot cause fungus. On the whole, the leaves generally show mixed infection of different fungal pathogen. Mainly old leaves show higher percentage of infection than the young one.

Management : Fallen and infected leaves collected and their destruction along with other sanitary measures should be periodically done to reduce the inoculums. Overcrowded branches should be pruned to ensure proper ventilation to reduce the conducive microclimate for disease development. As maintenance of orchard hygiene, through pruning of dead wood and removal of infected leaves twigs and fruit, helps to reduce the incidence of the disease, Spraying of fungicide Dithiocanbamate/Mancozeb/Chlorothalonil (@ 0.2 %) during fruit growth and just after rainy season also provides effective control against the disease. while pre-harvest applications of fungicides such as mancozeb and bezimidazoles offer control (McMillan, 1994a; Crane *et al.*, 1997).

Red rust (C.O. - *Cephaleuros virescens*)

The algal leaf spot of litchi was reported in India (Mishra *et al.*, 1974; Anu and Mandhar, 1994) and also from important litchi growing countries like China, Thailand (Menzel *et al.*, 1988; Alfiera *et al.*, 1994; McMillan, 1994b).

Litchi leaves, petioles and stem have been found affected by an algal parasite causing loss of vigour. It is prevalent in wet climates and thus generally appears during rains and continues to spread till early winter. The disease is characterized by red round misty spots on the leaves and some times also on the tender stems. The infection reduces the vitality of the tree by hampering photosynthetic activities. The growth of the plant is retarded due to defoliation which indirectly affects the yield. The velvety growth turns light brown to brick-red. The growth of this alga initiates the development of corky tissue in few upper layers of leaf tissue causing their death. Severely infected leaves exhibit curling inward towards dorsal side. Old and thick leaves show various types of malformation.

Management : Pruning and training of the dense canopy to improve tree ventilation. Spraying of Copper oxychloride @ 0.3% is highly effective in reducing the disease. The fungicides Captafol (0.2%) and Ziram (0.25%) have also been found effective to control the red rust.

Twig blight, leaf blight and die-back (*Lasiodiplodia theobromae (China)*; *Pestalotiopis mangiferae (India)*; *Armillaria mellea* (South Africa); *A. socialis* (Florida).

It start appearing brown spots on branches, followed by blistering and then cracking of the bark and ultimately death of the branch. The term die-back implies death of a plant from top downwards. Almost a decade back this disease was considered to be a minor disease but now the situation has changed. The disease is steadily gaining ground and spreading very fast into old orchards and causing rapid decline of litchi orchards in many litchi growing areas, especially in poorly drained flood hit soils. It can affect the whole tree or just one or two branches. The tree or branch may recover temporarily, but subsequently die. Disease development is favoured by rains, relative humidity (approx. 80%) and maximum and minimum temperature of 31.5 and 25^{0}C respectively. This fungus is also found to infect fruit and cause fruit rot.

Management : Pruning of dried/dead infected twigs/branches and pasting of cut ends suitable fungicides and also healthy bud wood with protective fungicides are the measures available to combat the disease. The orchard should be inspected regularly for any fresh infection to ensure disease free orchard by judicious pruning followed by spraying/pasting of Bordeaux mixture/Bordeaux paste. Two sprayings of Copper oxychloride (0.25%) + Carbendazim (0.15%) at 15 days interval are highly effective to combat the die-back of litchi.

Root rot (*Armillaria mellea*)

In some litchi-growing areas there are reports that trees are killed by the root rot. A slow decline and a sudden death have been recorded. It is soil borne disease. A loss of tree vigour and absence of new shoots are the first obvious symptoms of the disease (Darvas, 1992), which followed by leaf sheding and ultimately may die. Young trees may wilt suddenly but old tree decline takes few years to die. The symptoms include a sudden branch wilt that is followed by the decline of new growth on the affected branch over a period. The disease typically spreads in a circle in the orchard and it is more severe in stressed or malnutritioned trees. The fungus may survive in the soil, or on stumps and roots of various trees for many years.

Management : Carefully soil preparation for young plantation preferably soil treatment with methyl bromide. Infected and dried trees should be removed. Deep ploughing. Use of excess water in the litchi plantations must be avoided. Infected portion should be removed at possible extent. Soil drenching with fungicide such as Copper oxychloride (0.25%), Bordeaux mixture (1%), Carbendazim (0.1%) should be done. Antagonistic bio-control agents such as *Trichoderma* spp. should be applied at the time of establishing new litchi orchards.

Corky bark (*Fusarium decemcellulare*)

It is recently recognised disease of litchi in India, South Florida (Ploetz, 2003). The fungus is responsible for corcky bark. An armoured scale i.e. *Andaspsis punicae* has also been associated with corky bark lesions in litchi in Florida. Affected trees have small to large irregular patches of bark which become raised, cracked and rough. The bark surface in the infected area is friable and can be rubbed off easily. As these patches enlage and become more frequent, branch death occurs. Eventually, large portion of canopy can be killed.

Management : Removal and destruction of affected branches and tree orchard. Disinfection of pruning equipment to ensure that the pathogen is not spread. Use healthy planting material in new orchard (in no instance air layers taken from infected orchards affected tree).

Lichens

Certain lichens are very common on the trunks, branches and twigs of litchi trees. Generally, they donot assume serious proportion but if the plant is infested with lichens, it makes poor growth and reduces vigour. When such plants come to fruiting, they bear less and the aril development in fruits is very poor. Yield and quality are reduced drastically.

Management : The lichen infested trees can be invigourated by spraying with 1.0 % caustic soda (Chakravarti *et al.*, 1972).

Gummosis (*Botrydiplodia theobromae* Pat)

Though it is not prevalent in many areas but it is found occurring in the litchi growing mainly in sandy soils and poorly managed orchards. The infestation of the disease appear on main stem or its branches and visible just after the rainy season. It is characterised by oozing out of gum on the surface of the affected stem, bark of the trunk but found more common in cracked braches or the branches have bark splits.

Management : Gummosis in litchi is not found in the well managed orchards receiving regular sprays of copper fungicides. Bordeaux mixture pasting/brushing on the main trunk or its branches from ground level to atleast 1m height in commercial bearing orchards.

Fruit Yield

It has been found that the obtained yield from the old trees is high but fetching very less price in the market due to inferior quality particularly with respect to size and wastage due to attack of many physiological disorders and pests-diseases is very high. Fruit yield and physicochemical characters of mature fruits were found to be better in fruits obtained from rejuvenated trees came into bearing. Maturity period is found to be slightly delayed in rejuvenated plants.

Table 1. Fruit yield from unrejuvenated and rejuvenated trees

Particulars	Fruit yield (kg/tree)	Categorization of fruit under different quality classes			
		Extra Class	Class I	Class-II	Wastage
Old litchi tree	85	19	24	26	16
Rejuvenated litchi tree (third year)	28	12	10	04	02
(fourth year)	46	21	14	07	04

Economics

The various operations in rejuvenating old senile trees is not an immediate process, but usually it can be done effectively over a period of 2-3 years and later maintenance and management continues. The cost of reiterative pruning and subsequent operations in one acre of old senile orchard comes about Rupees twelve thousand two hundred and fifty (Rs. 12,250.00) only (Table 2). It has been found that the old senile orchard can be rejuvenated back just like young bearing orchard within a time gap of 2-3 years without any income loss in-between by adopting and following the technology, when the case was studied under close observation and monitored accordingly for entire activities and economical aspect was also taken care off (Table 3).

Table 2. Expenditure incurred in rejuvenating senile litchi orchard (One acre)

Operation	Amount (₹)
Charges for reiterative pruning 40 mandays @ ₹ 68.50 and 15 mandays @ ₹ 150.00 /day	4,990.00
Cost of ploughing, leveling and cleaning	750.00
Cost of manure and fertilizers	1,650.00
Cost of ring making and fertilizer application	1,050.00
Cost of irrigation	1,440.00
Cost of chemical pasting and spraying	1,110.00
Cost of thinning and other operations	960.00
Miscelaneous (Saw blade, rope, bamboo pieces etc.)	300.00
Total	**12,250.00**

Table 3. Economics of rejuvenation of old senile orchard (one acre) during 2003-06

Operation	2003	2004	2005	2006
Sale of fruits before rejuvenation	6500.00	-	-	-
Cost of rejuvenation	(-) 12250.00	-	-	-
Sale of woods	8,000.00	-	-	-
Cost of intercropping	(-)2,700.00	(-)4,500.00	(-)6,500.00	(-)6,000.00
Profit from intercrops	10,200.00	19,000.00	24,50000	22,000.00
Sale of fruits from rejuvenated trees	-	-	-	9,500.00
Gross income	**9,250.00**	**14,500.00**	**18,000.00**	**25,500.00**

Superior quality in different cases have been recorded from the rejuvenated trees (compared to un-pruned ones) in litchi (Kumar 2008).

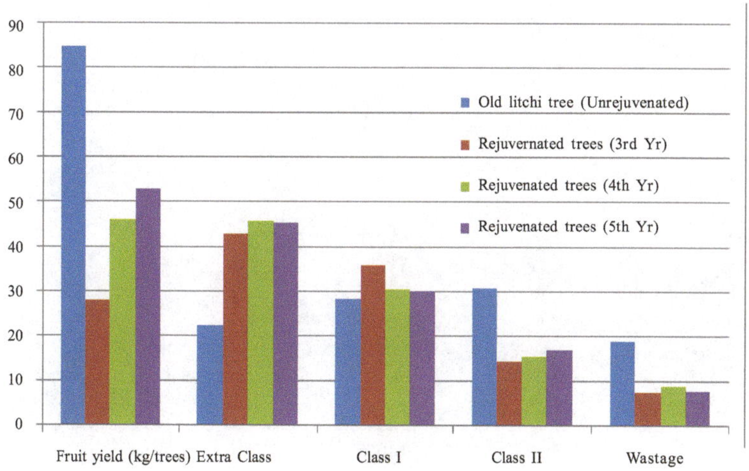

Fig 2. Fruit yield pattern of litchi cv. Shahi and its categorization of fruits under different quality class (%) in different rejuvenation technology treatments.

Source : Kumar, 2008.

Impact

The increasing trend of production and productivity of quality yield in coming years as commercial bearing orchards, the other litchi grower having old plantations giving uneconomical yield wanted to adopt the technology developed by the National Research Centre on Litchi, Muzaffarpur, and continuous approach is on to have necessary suggestions and guidance for rejuvenation to be done in their orchards. In India, the success of rejuvenation technology in turning the old senile orchards into young commercial bearing orchards, the National Horticulture Mission (NHM) of the state has also included this particular technology package in its action plan. The National Horticulture Board (NHB), Gurgaon has funded projects and National Banking for Agriculture & Rural Development (NBARD) have initiated programmes with enough subsidy to take this technology to farmers fields. The local news paper published and flashed this successful technology transfer with various success stories. The technology have been successfully transferred to almost all the litchi growing districts of Bihar and its spread is being made to other states too.

Old senile unproductive declining litchi orchard

Unproductive declining orchard

Neglected unproductive orchard

On-site training for rejuvenation

(A) Training to orchard owners for rejuvenation (A & B) (B)

(A) Reiterative pruning with the help of power operated chain saw (A & B) (B)

(A) Pruning with power operated chain saw (A & B) (B)

Measuring desired pruning height

Marking of tree for pruning height

Wrong method of cutting

Lopping for reiterative pruning

Reiterative pruning and sale of pruned woods

Logs of pruned woods for sale

Team for orchard floor management

Training to growers

Orchard floor cleaning

Reiteratively pruned orchard

Removal of unwanted part of trunk

Hard core pruned tree

Bordeaux paste on cut portion

Ring method for nutrient application

Cleaning of orchard floor

Lime brushing on pruned trunk

Emergence of sprouts on pruned tip

(A) Profuse emergence shoot thinning of shoots in rejuvenated trees (A & B) (B)

(A) Intercropping in rejuvenated orchard 6 months after pruning (A & B) (B)

(A) Various intercrops at different stages in rejuvenated orchard (A & B) (B)

Pruning of branches

Training of branches

(A) Top-grafting in rejuvenated tree (A & B) (B)

Pruning and training for canopy development

Canopy development after one year

Intercropping in rejuvenated orchard

Aerial view of rejuvenated orchard

Orchard floor management in rejuvenated orchard

Canopy development after 2 years of pruning

Developed canopy after 2 years

Rejuvenated tree in full bearing

Chapter 15

Mango

Mangoes (*Mangifera indica* L.) belong to the family Anacardiaceae, has become naturalized and adapted throughout the tropics and subtropics. There are over 1,000 varieties throughout the world, which is a testament to their value to mankind. Popularly known as 'King of Fruits', mango is eaten throughout the world. Commercial mango cultivation is based on locally adapted and recommended cultivars. There is a great diversity in mango fruit which permits considerable manipulation of various purposes and markets such as juice, candy, chutney, amchur, pickles, jam, jelly, fresh fruit, canned and/or dried fruit etc. Given the multiple products, it is therefore a potential source of foreign exchange for a developing country as well as a potential source of employment for a considerable seasonal labour force.

PRESENT STATUS

India is the leading country sharing 42.1 % of the world mango production. Currently, it is cultivated in 22,97,000 ha with a total production of 1,51,88,000 tonnes. In India, all fruits share 29.25 % of total land and 31.13 % of total fruit production of which, 36 % of land and 20.3 % of production is under mango. Mango prefers a warm, frost-free climate with a well defined winter dry season. Rain and high humidity during flowering and fruit development reduces fruit yield. The tree generally flowers in mid- to late winter, while fruits mature during early to mid-summer. Mango trees are usually between 3-10 m (10–33 ft) tall but can reach up to 30 m (100 ft) in some forests. The canopy is evergreen with a generally spreading habit. The heavy canopy of the mango is a source of shelter and shade for both animals and human beings. It occupies a pre-eminent place among sub-tropical fruits grown in some states.

PRODUCTION CONSTRAINTS

Though, mango is one of the important commercial fruit crop of India, it is least professionally grown. Its yield are generally poor because the tree is huge and traditionally planted at a wide spacing based on the tree's ultimate size and mostly with negligible management practices. The low productivity is due to majority of plantations are planted with non-descript or unknown seedlings. Further, alternate bearing, mango malformation, fruit drop, insect pest and disease infestation, dense and overlapping canopies, declining phase with negligible maintenance are other constraints. A large number of mango orchards are becoming old and tree canopies either have gone dense and overcrowding or are attaining the declining phase with no new growth and leaf flushes for fruit production. Dominant vegetative phase can affect their reproductive phase especially in tropical regions, if not regulated properly. Hence, physiological disorders and several pests and diseases affect mango all through the year. The yield loss due to pest and diseases is more than 15-20 % during normal year and it goes 40-60 % in severe infection. Pests like stem borer and bark-borer cause severe damage. Old/unproductive mango orchards having dense canopy hinders the proper aeration and sunlight interception throughout the year, leading to various disease and pest infections.

Jackson (1980) described a close relationship between light interception, photosynthesis and yield in fruit orchards. There is always unfavourable effects of low light interception and utilization on productive physiology of fruit trees. Light is an important factor for flower bud induction, fruit size and colour development. It has a role in fruit development through carbohydrate synthesis. While, increased assimilates in the shoots is pre-requisite for flowering in mango (Chako and Ananthanarayn, 1982), failure of flowering in mango trees with dense canopies (Burondkar and Gunjate, 1991) and opening canopies through pruning (Madharavarao and Shanmugavelu, 1976) and in shoot retardation (Rameshwar, 1989) support indirectly the role of light in fruit bud formation of mango. Canopy architecture is a natural expression of the genetic make-up of the tree, genotypes vary in their canopy size and shape. Certain scion growth controlling rootstocks and interstocks (Kurian *et al.*, 1996; Valdivia, 1999) have been utilized to restrict the mango canopy size.

Need of Rejuvenation

A large number of orchards remain neglected and can no longer attain full potential of fruit production. There is need to enhance their production. However, in old declining and unproductive trees, pruning and canopy re-building are the only and ultimate way for transforming them into young juvenile productive phase. Mango trees response to pruning significantly as (Rao, 1971; Hackett, 1985; Lal *et al.*, 2000; Davenport, 2006) reported the increase in fruit yield in pruned mango trees than unpruned trees because of increase in photosynthetic surface and increased availability of light to newly-developed canopy. To rejuvenate the unproductive mango orchard and by proper pruning and training the canopy rebuilding for increased surface bearing area for enhanced quantum of quality production in sustainable manner. The pruning of mango is also not recommended to be followed as regular cultural operation, but it is to be applied under special circumstances. The majority of plantations that are of seedling origin are poor in yield, grown old/unproductive and senile. By adopting rejuvenation technique these seedling trees can be replaced with location-specific commercial varieties by top working technique. Therefore, focused programme should be shifted from replacement or area expansion to rejuvenation of old over-crowded, senile unproductive orchards.

Cutting down the unproductive orchard and establishment of a new one takes around 10 years to attain the commercial bearing stage. The technology has been developed to increase the productivity of such type of orchards through rejuvenation and canopy rebuilding. The canopy architecture scheme even under medium-density of planting in mango is to go for precision farming with controlled canopy where all the operations can be monitored efficiently, have been found successful. The approach of proper canopy architecture will also offer a way to adopt improved and efficient technologies. The open canopy can be made of mango tree by rejuvenating. It lowers down the disease pest infestation.

REJUVENATION TECHNIQUE

Identification of Old Senile Orchards

India is the largest mango producer in the world but commercial cultivation is not encouraging. Till date many orchards have not received proper cultivation practices. A considerable area under mango orchards has either gone unproductive or showing marked decline in productivity. The orchards (even of seedling origin) with unmanageable canopy do not produce good yield. The over-crowded and intermingled branches allow poor availability of light to growing shoots within the canopy. Thus, lanky growth and meagre foliage show die-back symptoms. The trees of seedling origin particularly showing dominance of vegetative growth can be selected for pruning and followed by top working for desired outcome.

Training to Orchard Growers

Rejuvenation technology is invasive in nature and time taking procedure. Thereafter, pruned trees also need intensive care and management for survival after-shock caused by severe pruning otherwise high mortality occurs. Hence proper training for creating awareness, interest to adopt the technology by showing its economic feasibility / success stories are required. Proper demonstration of the entire skilful operations through on-site (theory and practical) training is highly beneficial and effective.

Technique

In old senile unproductive robust trees of mango, operations like hard core pruning or pruning and top working for rejuvenation are required to be performed in scientific and skilful manner with utmost care in phases (Burrondkar *et al.*, 2000; Chandra *et al.*, 2012).

First phase

November, unproductive senile trees should be selected and marked. Four to five outward growing primary branches should be selected and marked so that after head back tree would form open type umbrella like canopy. Inward growing intermingle branches are selected for removing. In December, when plants go in vegetative dormancy, selected primary branches should be headed back at a height of 5 m above the ground level, whereas, undesired branches should be removed from the base of branch.

Pruning should be initiated from lower surface of the branch and alter from upper surface to avoid cracking of branch and bark splitting. Copper oxychloride paste (dissolve 2 kg of copper sulphate in 15 litres of water; mix 3 kg of quick lime in 2-3 litres of water and then add the remaining 12-13 litres of water; mix the two concentrated solutions) should be applied on cut surfaces as well as the trunk to check microbial infection (Lal *et al.*, 2000). Irrigate the orchard at regular intervals as per the requirement which will help in sprouting of new growth. Numerous branches will emerge from the headed back limbs.

There are four techniques for rejuvenation of old and senile mango orchards developed at different institutions in India and abroad which are as under:

Technique 1 : In this technique, the tall central trunks are cut back to about 3-4 m height from the ground. The actual site to cut back should be at a point where there are side branches. A chain saw is necessary for the job to make a smooth cut. Here, one half of the tree is cut back. The remaining trunks and leaves help protect the stump from sunburn. The freshly exposed trunks and branches to the sun should be painted with white water-based paint diluted 3 or 4 times. This is to prevent sunburn, which could attract borer on to the damaged bark. In a short time, numerous shoots will develop; select the most vigorous ones, spaced evenly around the stump and if possible at differing heights. Remove all the unwanted shoots. This process should be repeated as often as necessary until the selected shoots begin to dominate and take over. Then, the rest of the tree can be cut back and the operations may be repeated. By this method, half the yield will be realized throughout the process till it regains the vigour.

Technique 2 : Old, unthrifty trees can often be rejuvenated by a moderate to severe pruning. This is in the form of skeletonising the tree, i.e. cutting back the branches of the tree till only the basic frame is left. Moderate skeletonising would entail cutting back the large branches to healthy wood, while at the same time maintaining the basic structure of the tree. Particularly large unthrifty trees would benefit from a more severe skeletonising, where all the main branches are cut back to the main trunk only, leaving about a metre or so of branch. The whole trunk and remaining branches should be painted with diluted paint to prevent sunburn and borer attack. In both these cases, there will be a proliferation of sucker growth. These should be properly thinned out.

Technique 3 : In old and dense mango orchards, light interception and photosynthetic potentials of trees is reduced resulting in poor yield. The branches existing on main trunk are considered as first order branches, the branches existing on first order branch are called second order branches; the branches existing on second order branch are called third order branches, similarly fourth and fifth order ones. At IIHR, Bangalore, pruning third order branches 30 cm from the point of origin recorded the maximum yield in Alphonso (86.3 kg/tree). The trial conducted at CISH, Lucknow, revealed that pruning second order branches recorded maximum pooled fruit yield in Dashehari mango after 12 years of pruning (57.99 kg/tree).

Technique 4 : The Central Institute of Subtropical Horticulture (CISH), Lucknow had conducted a trial to rejuvenate 40-50 years old mango trees at farmers fields. In this process, main branches were pruned at a height of 5 m from the ground level in December. About 3-4 diverging branches were kept for developing healthy umbrella-shaped canopy, while rests were removed from the base. Pruned surfaces were smeared with copper oxychloride paste immediately after pruning to check the microbial infection. Pruned trees were kept under intensive care and management. Cultural practices like nutrition, irrigation, hoeing, weeding were followed properly. Profuse shoots emerged from prune branches from April onwards. Only 8-10 outward growing well spaced healthy shoots were retained per branch, while rest were removed. Plant-protection measures were seriously adopted especially against stem-borer, leaf cutting weevil and anthracnose because pruned trees came into flowering and fruiting 2 years after pruning. The growers ended up with loss by missing crops for 2 year, hence, technology was refined to undertake the pruning in alternate rows. With alternate row pruning, availability of light to unpruned trees in two adjacent rows was greatly improved and their fruiting increased by 2-3 times. Thus, enhanced production from unpruned trees compensated the loss to some extent. But for uniform result with synchronous flowering and

Old senile unproductive mango tree

Pruning in steps

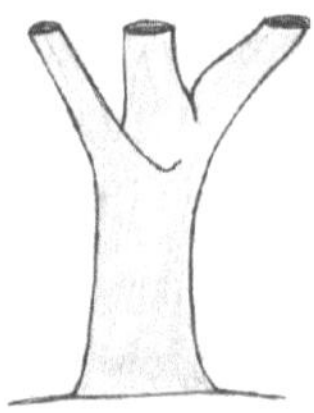

Pruned tree

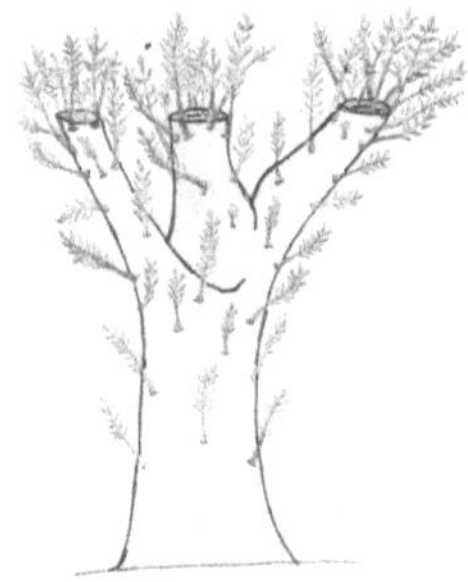

Emergence of shoots

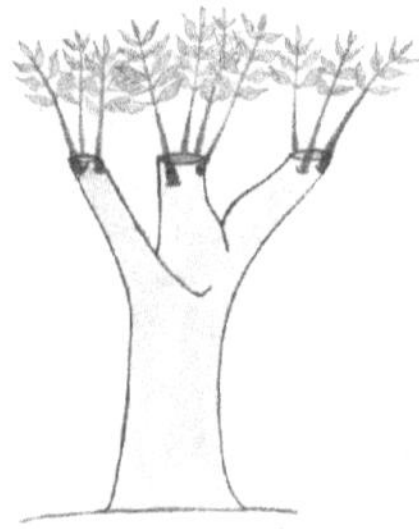

Shoot thinning

Rejuvenated tree in fruit bearing

fruiting, block/plot wise operations should be practiced as found beneficial in case of litchi (Kumar, 2008).

Sanitation and Aftercare

Clean the orchard from the pruned wood and plough it. Basins and irrigation channels should be prepared. Mulching should be done around the basins. Full dose of single super phosphate (3.00 kg/ tree) and half dose of urea (1.25 kg/ tree) in basins should be applied during February end. Careful observation for infestation of stem borer insect pest in pruned trees should be taken. Upon identification of infestation, placing cotton wick soaked or Dichlorphos or kerosene oil or inject water emulsion of 0.05 % Monocrotophos or Chloropyriphos. Irrigation should be given as per the requirement. Care of new emerging shoots should be taken.

Second phase

During June-July, undesired shoots should be removed while about 8-10 healthy shoots with outwardly growth per pruned branch should retained. The diseased, week and upward growing branches should be removed. Spray of copper oxychloride (3 g/litre) on the pruned trees and application of copper oxychloride paste on cuts. Apply remaining half dose of urea *i.e.* 1.25 kg/tree and FYM (120 kg/tree) in June in the basins. Plant should be protected from the incidence of different diseases and pests.

Third phase

Pruned trees need intensive care and management practices for survival after a shock caused by severe pruning. Proper care should be taken for the selected and healthy shoots until the canopy develops (Hassan *et al.*, 2009; Jahangeer *et al.*, 2011). In third year application of Paclobutrazol at the rate of 5.50 and 8.25 g a.i./tree both as a soil drench and spray applications were effective in suppressing vegetative growth and enhancing the flower growth. The increased percentages of flowered branches for Paclobutrazol treated trees is due to lower expenditure of tree reserves to the vegetative growth parameters and consequently no assimilate limitations (Davenport, 1993).

Top-working in inferior mango trees

Grafting and budding : Grafting is used to join the upper part (the scion) of a selected variety to the lower part of another (the rootstock)—the upper part is an identical clone of the desired tree. Almost all methods of grafting can be adapted for mango. Two popular methods for mangoes are the cleft graft and the whip-and-tongue graft. To achieve a successful graft, it is important to

Old senile unproductive mango tree

Pruning in steps

Pruned tree

Emergence of shoots

Top grafting on selected shoots

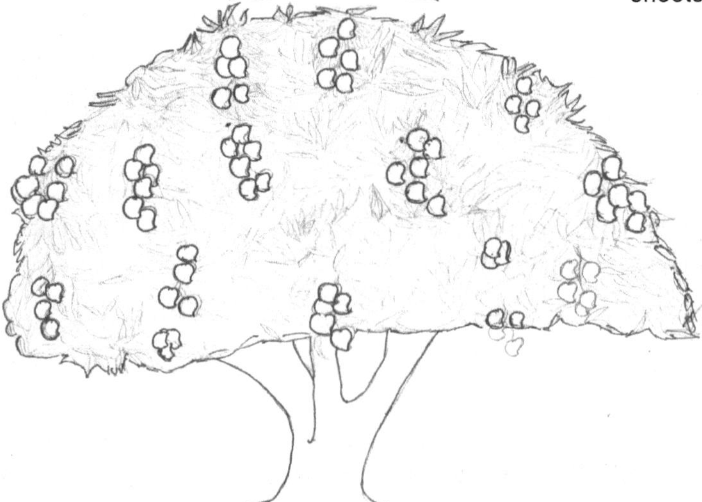

Rejuvenated (top grafted) tree in fruit bearing

have healthy, actively growing new branch and select scion wood with swollen buds that are ready to burst. It is also important to match the cambium layers in both the stock and scion, as the cambium is where the cells are actively dividing and the joining of the graft takes place. This is most easily done when the stock and scion wood are of the same diameter. In mango, the cambium layer is the white woody layer just below the bark. In the wedge method, the scion is prepared by making two sloping cuts that form a V shape. A straight cut is made on the stock of the same depth as the V shape on the scion. The V-shaped scion is then inserted into the stock, making sure that the cambium layers of both match up. The whole joint is then wrapped with grafting tape to hold it in place, and a plastic bag is used to cover the scion and union to prevent excessive moisture loss during the healing process. The whip-and-tongue method is similar to the wedge but uses interlocking cuts that give a stronger graft and greater surface area for healing of the graft.

Scion selection : Scion wood should only be collected from actively growing trees that show no symptoms of disease infection. Scion wood is best taken from terminal shoots with swollen terminal buds about to burst. The scion thickness should match the stock thickness to which it is to be grafted (6–15 mm) diameter and be between 100 and 200 mm long. After selecting and removing an appropriate piece of scion wood from the tree, all leaves should be immediately removed to reduce moisture loss through transpiration. The scion should be placed in a plastic bag and held in a cooled box at 5–10°C. If the scion wood is not be used within 12 hours of collection, it should be wrapped in moist paper toweling inside a plastic bag and stored in the vegetable crisping section of a domestic refrigerator. Scion wood can be kept in this way up to 2 weeks.

Pre-grafting : Scion budwood should be dipped in a fungicide and insecticide solution to prevent the spread of insect, pests and disease. Management practices to be followed by mango growers in trees pruned/headed back for rejuvenation.

- Preparation of basin around trees for proper manuring, fertilization and regular irrigation.
- Apply recommended dozes of urea, di- ammonium phosphate and muriate of potash along with 50 kg/tree of well rotten Farm Yard Manure.
- Apply copper oxychloride paste on the cut ends, whereas, on the large limbs apply paste of copper oxychloride and lime to prevent sunburn injury.
- Irrigate the plants soon after rejuvenation for shoot sprouting and for proper development of tree canopy.

- Pull out the trunk borers mechanically with the help of iron spoke in the shelter holes.
- Use black polythene for mulching (100 micron) soon after rejuvenation.
- Observe incidence of stem borer. In case of infestation, measures, comprising pulling out the grubs from the holes with help of iron spoke and plugging the holes with Monocrotophos soaked cotton swabs following mud plaster to be taken up.
- Irrigate the plants soon after rejuvenation for shoot sprouting and for proper development of tree canopy.
- Pull out the trunk borers mechanically with the help of iron spoke in the shelter holes.

Orchard Floor Management

The chief economic consideration, which encourages growers to go for mango rejuvenation is that, the tree does not suffer much, if it is not watered during hot months. Adequate moisture is required after (15-20 days) heading back of branches for proper initiation and development of shoot growth. Care of pruned tree requires watering them regularly during the dry season. Due to moisture stress in the pruned trees, emergence of new shoots as well as rejuvenation process is severely affected. Therefore, it is necessary to ensure irrigation in rejuvenated trees which is required at regular intervals for initiation of shoots below the cut portion. To promote the proper development of tree canopies and fruiting twigs, irrigation required at the interval of every seven to ten days in summer and fifteen to twenty days in winter in addition to the period of rainfall during the monsoon season.

Integrated Nutrient Managment

Integrated nutrient management in mango refers to maintenance of soil fertility and plant nutrient supply to an optimum level for sustaining the desired crop productivity in rejuvenated orchards through optimization of benefits from all the possible sources. Therefore, it is a holistic approach where we first know what exactly is required by plants for optimum level of production. In what different forms these nutrients can be applied in soil, at what different timings is best possible method, and how best these forms can be utilized properly by the plants. The amount of fertilizer to be applied, depends upon the age of tree, condition of tree and type of soil. For proper growth and profitable yield, fertilizer should be in the required optimum dosage. An adult bearing tree needs - 120 kg well rotten FYM, 4 kg Neem cake, 2.50kg urea, 3.0 kg single super phosphate and 1.5 kg muriate of potash. It should be applied preferably in two split doses, one during February end and the remaining half dose in June (Kumar and Adak, 2013).

Water Management

Water management is integrated to obtain higher productivity with efficiency of economically acceptable limits in environmental friendly way. In rejuvenated mango trees, along with nutrient, water management assumes much more significance. These two inputs are essentially required to be managed in a manner, which provides maximum output. Irrigation should be applied round the year but more frequently during summer months so that actively growing vegetative shoots never experience stress condition (Larson *et al.*, 1989).

Mulching

Mulching at the base of pruned trees is done by using black polythene sheets (100 micron or 400 gauges) or heavy mulching with organic material, such as straw, dried grass, banana leaves surrounding the main trunk. Mulching with the organic material should be applied thick enough (12-15 cm) to prevent weed growth yet permit rain water penetration to the root area. Black polythene sheets prevent soil surface evaporation and tend to produce water under the sheets through condensation, supporting tree growth besides checking weed growth. There is considerable reduction in water application in polythene mulched orchards as compared to unmulched orchards. Reduction in water could lead economized inputs and enhanced production per unit area (Kumar 2015).

Intercropping

Regular intercultural practices are essential for proper upkeep of the pruned mango orchards. The practice of intercropping improves physical conditions of soil, ensures aeration by breaking soil surface crust and removes those weeds which compete for soil moisture and nutrition. In order to manage the orchard soil, two time ploughing is necessary in the year. One ploughing should be done in June while the second ploughing should be done in the month of December. The first ploughing helps in checking the run off losses, facilitating maximum intake of water into the soil. The second ploughing checks the weed growth and induce vegetative shoots.

Intercropping is intended to maximize land and space-use efficiency to generate supplemental income, particularly during the phase of canopy development in pruned orchards. After rejuvenating the mango orchard, tree takes about 2-3 years to develop canopy and cover the area. However, care should be taken in selecting the right type of intercrops within the rows of pruned mango trees. Vegetables and leguminous crops can easily be grown up to three years after rejuvenation. Cowpea, beans, cabbage, cauliflower, chillies, okra and partial shade-loving crops (ginger, turmeric and elephant foot yam) as an intercrop in the orchard provide suitable return from the initial stage of canopy development (Kalloo *et al.*, 2005, Alila, 2012).

Managing Physiological Disorders

Fruit drop

The natural fruit drop of mango is rather amounting to about 99% at various stages of growth. Regular irrigation during the fruit setting and development period can reduce fruit drop considerably. Application of plant bio regulators like NAA (40 ppm) or 2,4-D(20 ppm) - about 6 weeks after fruit set, reduce fruit drop considerably. Spray of 0.8% zinc sulphate or 4.0% potassium nitrate at bloomstage is very effective in increasing the fruit retention, yield, quality and shelf life of mango cv. Dashehari.

Biennial bearing

This is one of the most burning problems it renders mango cultivation less remunerative to the growers. Biennial bearing is also known as alternate bearing. It indicates yield variations in alternate years *i.e.* an year of optimum or heavy fruiting followed by a year of little or no fruiting. The problem of biennial bearing is a varietal character governed by genetic makeup, and this tendency starts exhibiting in mango plant even at the second year of fruiting and become more serious as the age advances. Immediately after harvesting, irrigation and nutrition management can reduce this problem.

Mango malformation

Mango malformation is of two types vegetative and floral. The vegetative malformation generally affects seedlings of young plants in which there is a swelling of buds and formation of small shoots with short internodes at the apical portion and given an appearance of witches broom like structure. Infloral malformation, panicles become deformed, axis become short and rachis thick, due to this inflorescence look like a cluster. Malformed panicles have bigger flowers than the normal flowers and are mostly male.

For reducing this malady, the first or early bloom is discarded. Affected malformed shoots should be removed and destroyed. The next blooms produced are of normal type and set fruits. De-blossoming with spray of 200 ppm NAA is helpful to some extent (Majumdar *et al.*, 1970; Haldankar *et al.*, 2013).

Leaf scorch

Mango leaves particularly old one show scorching at the tip and margins. Affected leaves fall down and the tree vigour and yield are reduced. The main cause of this malady is due to an excess of chloride ions which render potash unavailable. Proper and timely irrigation and nutrition management can reduce the occurrence of this disorder.

Black tip

Black tip disorder is widespread in India. It may the first symptom of this disorder is etiolation and yellowing of the distal end of fruit. Later affected fruit show brown to black necrosis of tissue, which may hardened at the distal end. The affected areas are flattened and sunken, and the inner portion is soft with dark brown liquid oozing from the tissue.

The practice of spraying three times with borax (0.6%) and caustic soda (0.8%) at stages (i) prior to flowering (ii) during flowering/bloom and (iii) at fruit set (Majumdar and Sharma, 1985) can take care from this disorder.

Mango decline

Mango decline is becoming a serious problem in major parts of northern and central India. Old plants show the symptoms of declining trend invariably. Initially, the symptoms appear on older leaves with collapsing of tissues. Besides, striking irregular brown necrotic area may also develop near the leaf tip and margin. Later, with increasing severity almost entire lamina gets affected. It gradually leads to the mortality of upper twigs progressing further larger branches. Affected trees have thin canopies of lustreless green leaves and sickly look. Decline is associated with several biotic and abiotic stresses. The poor orchard floor management and poor nutrition are the major cause of the decline. Salt injury is also one of the main cause of the decline in many areas. The rejuvenation technology is only adopted or feasible if planned after managing to get rid of this malady.

Growing of dhaincha as green manure in the interspace of the orchard is quite helpful. Apply gypsum (50 kg / tree at >50 years aged) around the basin beneath the canopy. The farmyard manure and compost should be applied every year in the orchard. Reduce the salt concentration if any to nontoxic level, by heavy irrigation with fresh water. Spray of micronutrients to avoid the deficiency (mainly boron and zinc).

Management of Pests

Mango hopper (*Amritodus atkinsoni*; *Idioscopus clypealis*)

The hoppers are widely distributed and serious pests of the entire mango-growing regions in India. There are two main types: *Amritodus atkinsoni* is dominant in north part of India while *Idioscopus clypealis* is more serious in southern part. The pest remains active throughout the year but maximum damage is inflicted during February-April. During extreme climate only adults are met with. The damage is inflicted by both nymphs and adults through sucking cell sap. The nymphs are found clustering on the inflorescence and suck the cell sap in spring season. The infested flowers shrivel, turn brown and ultimately fall off. These hoppers also excrete honeydew, which encourages the growth of sooty mould on leaves, branches and even on fruits.

Proper pruning and training for good canopy development and more light interception. Spraying of insecticide well before the flower opening on panicle is also good. Spray Dimethoate (30%) @ 1ml/litre or Cypermethrin (25%) @ 0.5 ml/litre for effective control.

Mealy bug (*Drosicha mangiferae*)

It is a multiphagous pest, it is widely distributed in India. The damage starts soon after the hatching, as nymphs start crawling up the tree trunks and clusters of population on various twigs, panicles and fruits found sucking the sap. Only nymphs are destructive and cause the tender shoots and flowers dry up. The young fruits also become juiceless and drop off pre-maturely. During heavy attack, the trees retain absolutely no fruit.

Dig or plough around mango trees as intercultural operation during summer to kill hibernating eggs. Spray of Imidachlorpid (0.5ml/litre) on entire body of the tree twice at the interval of 5-7 days.

Stem-borer (*Batocera rufomaculata*)

The beetles have been recorded as serious pests of mango. The borer is not very common, yet when it appears in the main trunk of a branch, it generally kills the host. On hatching, grubs make zig-zag burrows beneath the bark and tunnel into the trunks or main stems and move upwards and continue feeding on the internal tissues. The attack of the pest is known from the frass that comes out of the entrance hole made by the grub. The adult beetles damage by feeding on the bark of young twigs and petioles. The pest causes considerable damage during March-April.

Removal/clearing of frass followed by pluging the holes with inserting the Nuvan (2ml/litre) soaked cotton balls into the holes and sealing the opening with mud is advocated to control the pest.

Shoot-borer (*Chlumetia transversa*)

Mango shoot-borer has also attained the status of pest in India. Its attack is more pronounced in nurseries and young grafted seedlings. Freshly hatched caterpillars bore into midribs of tender leaves or the tender shoots near the growing point, tunnelling downwards, leaving excreta outside the entrance hole. Leaves of affected shoots wither and droop down and conspicuous from a distance. The pest is active from August to October.

Removal and burning of the dried up shoots reduces population pressure of pest. Spray of Carbaryl @ 2g/litre or Dimethoate @ 1ml/litre on emergence of new vegetative flush or flowering panicle has been found effective.

Mango scales (*Aspidiotus destructor*)

The scale insect becoming serious pest of mango. It is found active during summer and affects the leaves and fruit setting adversely. Nymphs generally suck the sap from the leaves, twigs and occasionally from the fruits. The chlorophyll of the attacked portion is destroyed and attacked spots appears yellow. In severe infestation, sooty mould develops, which greatly retard the photosynthetic activity of plants. The scales can be easily removed from the attacked part of the tree.

Spray of 3-4 ml/litre of Chlorpyriphos in March and September can control the infestation of scale insects.

Management of Diseases

Powdery mildew (*Oidium mangiferae*)

Powdery mildew has become limiting factors in getting good yield in mango. It gests very serious after flowering. Whitish or greyish powdery mass appears on young foliage and inflorescence. The affected panicles bear very little or no fruit due to pre-mature withering of inflorescence or fruit drop. Warm humid weather and low night temperature favour disease development. The fungus perpetuates on the infected trees.

The disease can be controlled by spraying Karathane (0.1 % or wettable sulphur (0.25%). Spray once before flowering and again during flowering and then after fruit setting are essential.

Anthracnose (*Colletotrichum gloeosporioides*)

The disease appears on young, leaves, shoots inflorescence and fruits. Round or irregular brown spots appear on leaves and young shoots. Spots on leaves enlarge with cracked centre. Spots on shoots enlarge longitudinally giving a blighted appearance. In severe attack, the shoots may die. Minute black dots appear on inflorescence and fruit skin. Affected fruit surface becomes sunken. The optimum temperature for the development of anthracnose is 25°C. The disease spreads rapidly during rainy season. Cloudy and misty weather during flowering accelerates damage to the infected floral parts.

Prune and destroy the shoots showing spots of canker and dead branches. Apply Bordeaux paste and then paint to the cut ends. Protective sprays with Bordeaux mixture (2:2:250) or copper oxychloride (0.3%) help in reducing the spread of the disease.

Twig die-back and leaf blight (*Macrophomina mangifera*)

Die back is one of the serious diseases affecting mango plantation and may be the cause of decline in commercial bearing orchards. The disease is characterized by the death of twigs from the top downwards particularly in the older trees followed by complete defoliation. Discolouration and darkening of the bark on or at a certain distance from the tip is external evidence of the disease. The bark of the dead twigs become completely discoloured and shrivelled. Cracks may also appear in partially dead bark of the affected twigs from where gum may exude before death. Initially, minute, yellow spots turning brown appear on leaves and twigs. The spots increase gradually and cover a large leaf surface. The spots have dark purple margins. Dark brown lesions appear on young and ripened fruits.

The disease is more conspicuous during October-November. Moderate temperature (26-32°C) coupled with high relative humidity and rains favour the development of the disease. Infected grafts, twig bark of the trees are source of primary infection.

The preventive measures may be the selection of scion from healthy trees and adoption of sanitary measures at nursery level and good orchard floor management with proper plant protection measures (IPM) *etc.* Pruning and destruction of diseased twigs upto 5 cm below the site of infection and protecting the cut surfaces with Bordeaux paste followed by spray of Bordeaux mixture effectively checks the disease.

Stem canker (*Schizophyllum commune*)

It is a major disease problem of old and neglected orchards. Cankers are formed on the stem which leads to discolouration and drying of foliage of one or more branches. Gum exudation from the affected branches is common. Branches dried in due course. Small shell like dirty white fruiting bodies of the fungus appear in rows on the dead portion of main branches with gills on the lower side. The pathogen has wide host range including several fruit and forest trees. The fruiting bodies are produced in abundance after moist conditions. The fungus gains entry through wounds on tree trunk.

Spray trees with copper oxychloride 50 WP @ 3 % or Bordeaux mixture (2:2:250) at fortnightly intervals. Spray particularly if untimely rains are received. Apply Bordeaux paste at wound portion after removing the disease parts.

Bacterial canker (*Xanthomonas campestris* pv.*Mangiferae*)

Minute, dark green, water-soaked areas appear on leaves, leaf stalks, stem, twigs, branches and fruits. Later, these spots become raised. These spots on fruits grow bigger usually up to 5 mm in diameter, which cover almost the whole fruit. Later these often burst extruding gummy substances containing highly contagious bacterial cells. Sometimes, the exposed flesh cankerous spots attract insects, subsequently paving the way for secondary infection, which initiate rotting. The disease spread is rapid during rains and becomes severe in July-August.

Regular inspection of orchards, sanitation and seedling certification are recommended as preventive measures. Orchard sanitation, use of windbreaks and sprays of Streptocycline (200-300 ppm) followed by copper oxychloride (0.3%) at 20 days intervals are found effective.

Gummosis (*Lasiodiplodia theobromae* Pat synonym : *Botrydiplodia theobromae* Pat)

Gummosis is more prevalent in even young orchards when established in light soils but it is found occuring in other type of soils also. The disease appear on stems and branches after rainy season.

Removal of diseased bark portion and pasting with Bordeaux paste or copper oxychloride help in keeping the disease under control. Application of copper sulphate @ 500 g in soil around the tree base is recommended practice for management of the disease (Singh *et al.*, 2008).

Red Rust (*Cephaleuros viresens* Kunze)

Rusty red fructification of alga on the surface of the leaves, veins, petiole and young twigs. Initially greyish grey colour circular spots velvety in texture appear usually on old leaves and rarely on petioles and young leaves. The spots then become irregular and finally turn reddish brown. The fruiting bodies are formed in moist atmosphere and the disease is more prevalent in dense orchards and act as parasitic way and ill effect is more pronounced in weak trees.

Canopy management, proper intercultural operation beneath the tree canopy, timely application of manures and fertilizers can get rid of the pest infestation. Spraying of copper oxychloride (0.3%) or algaecides (Fentin acetate) have been found effective in controlling the infestation.

Lichens (*Strigula elegans* (Fee) Mull Arg.)

Lichens are found on fully grown trees of mango, mainly on trunks, branches and twigs in areas of high humidity, heavy rainfall and poorly managed orchards. Lichens are seen in the form of whitish, pinkish superficial coverings of different shapes on the main trunk, branches and twigs of trees. In general, trees with severe lichen infestation show poor growth. Lichens donot damage the plant body directly but give them the ugly look and unhealthy appearance and bring about growth reduction by disrupting the normal physiological process of the tree.

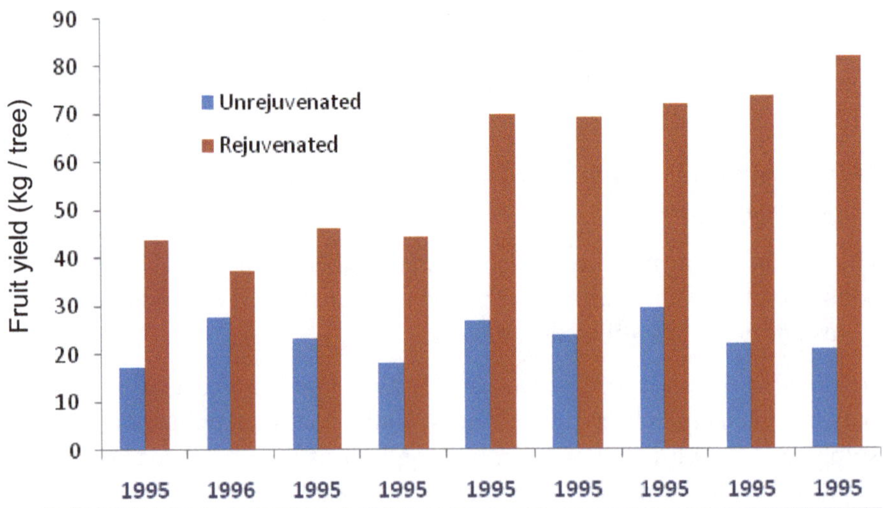

Fig 1. Fruit yield pattern of rejuvenated Dasheri trees

Source : Kalloo *et al.* 2005

The lichens can be controlled by gunny rubbing followed by spraying the trunk, branches and twigs with commercial caustic soda (1.0%) and then followed by spray of copper oxychloride (3.0%) effectively control lichens.

FRUIT YIELD

In mango, fruit production starts from the third year and the productivity increases upto about four times after proper canopy development of about 7-8 year of the operation. Thus, productivity is restored and with progressive increase in fruit yield making the orchard productive and remunerative (Kalloo *et al.*, 2005; Lal *et al.*, 2000, 2001; Lal and Mishra, 2007, 2008; Reddy and Kurian, 2011; Kumar and Kavino, 2012; Mishra 2013). An enhancement of about 22.35% to 45.65% in fruit yield has been recorded.

Economics

The various operations in rejuvenating old senile trees is not an immediate process, but usually it can be done effectively over a period of 2-3 years and later maintenance and management continues. The cost of reiterative pruning and subsequent operations in one acre of old senile orchard comes about Rupees twenty six thousand eight hundred and fifty (Rs. 26,850.00) only (Table 2). It has been found that the old senile orchard can be rejuvenated back just like young bearing orchard within a time gap of 2-3 years without any income loss in-between by adopting and following the technology, when the case was studied under close observation and monitored accordingly for entire activities and economical aspect was also taken care off for the economic feasibility (Table 3).

Table 1. Expenditure incurred in rejuvenating senile mango orchard (One acre)

Items/Operation	Amount(₹)
Charges for reiterative pruning 40 mandays @ ₹ 200.00 and 15 mandays @ ₹ 300.00 /day	12,500.00
Cost of ploughing, leveling and cleaning	2,000.00
Cost of manure and fertilizers	2,250.00
Cost of ring making and fertilizer application	1,600.00
Cost of irrigation	3,000.00
Cost of chemical pasting and spraying	2,500.00
Cost of thinning and other operations	2,000.00
Miscelaneous (Saw blade, rope, bamboo pieces etc.)	1,000.00
Total	**26,850.00**

Table 2. Economics (₹) of rejuvenation of old senile orchard (one acre) during 2004-06

Items/particulars of operations	2004	2005	2006	2007
Sale of fruits before rejuvenation	15000.00	-	-	-
Cost of Rejuvenation	(-) 26850.00	-	-	-
Sale of woods	60,000.00	-	-	-
Cost of Intercropping	(-)4,500.00	(-)7,000.00	(-)7,500.00	(-)10,000.00
Profit from Intercrops	12000.00	20,000.00	25,000.00	30,000.00
Sale of fruits from rejuvenated trees	-	-	-	40,000.00
Gross Income	**55,650.00**	**13,000.00**	**17,500.00**	**60,000.00**

Source : Farmers orchard (Mr.BS Randhawa), Ramnagar, Uttarakhnad.

CALENDAR FOR REJUVENATION

Year 1

December - January

- Marking of trees and their undesired branches for pruning.
- Pruning of marked branches in December.
- Pruning to be followed in half portion or entire orchard.
- Pruning to be initiated after lopping off operation to skeletonise the tree.
- Reiterative pruning taking lower surface of the branch making small logs.
- Pruning carefully taking alter from upper surface to avoid cracking of branch and bark splitting.
- Final cutting by way of reiterative pruning at desired height from the ground level.
- Application of copper oxychloride paste or biodynamic tree paste on the trunk, branches as well as cut surfaces to check microbial infection.
- Sharp pruning and cleaning of the top portion from dust or rough surface to prevent ascent of nymphs of mealy bug.
- Preparation of basins and irrigation channels after a month gap.

February - March

- Ploughing and weeding of orchards in after a month gap of pruning operation.

- Application of recommended full dose of single super phosphate (3.00 kg/ tree and half dose of urea (1.25 kg per tree) in basins by the February end.
- Careful observation for infestation of stem borer insect pest in pruned trees.
- Upon identification of infestation, placing cotton wick soaked with Dichlorvos or kerosene oil or water emulsion of 0.05% Monocrotophos.
- Spray Carbendazim 50 g /100 liter of water or wetable Sulphur 200-250g / 100 litres of water Spraying against hopper

April - May

- Irrigation as per requirement.
- Mulching in basins around trees.
- Hoeing and weeding in basins.
- Care for new emerging shoots.
- Observation for incidence of stem-borer and its management.

June - July

- Thinning out of undesired shoots while retaining about 8-12 healthy shoots with outwardly growth per pruned branch in June followed by spray of copper oxychloride 3 g/litre.
- Irrigation at an interval of 10-15 days.
- Application of remaining half dose of urea. i.e. 1.25 kg / tree during June.
- Application of FYM (120 kg/ tree) in basins during July.
- Management of stem borer accordingly.
- Spray of copper oxychloride (3 g/litre water) twice at an interval of 15 days, if there is infestation of anthracnose and other leaf spot diseases on new leaves.
- If there is serious incidence of leaf cutting weevil, two sprays of 2% Carbaryl (Sevin) @ 2 g / litre water at an interval of 15 days may be done.
- Sowing of green manuring crops or rainy season inter crops.

August - September

- Thinning out undesired shoots.
- Observation of incidence of stem-borer insect pest and anthracnose and other leaf spot diseases and their management.
- If attack of mango leaf-webber is noticed spraying 0.04 % Monocrotophos is recommended.
- Repeat the spray for control of scale insects, leaf, shoot borers and leaf cuttings weevils as suggested in previous month.
- Ploughing of green manuring crops.
- Ploughing and cleaning and removal of weeds to be done and leaf cuttings weevils.

October - November

- Cultural operations of ploughing, hoeing, weeding etc.
- Removal of dried and diseased twigs.
- Management of insect pests and diseases.
- Foliar spray of 2% urea during October for healthy vegetative growth.
- Marking of trees for pruning.
- Spray 200 ppm naphthalene acetic acid in the 2 fortnight may be done to overcome the mango malformation.
- Mixing of 75 –100 kg well-decomposed farmyard manure in the basins of each mango plant.
- To control die back disease, cut the affected twigs and burn.
- Spray 0.3% copper oxychloride, and the spraying may be repeated after 15 days. If required, repeat the spray for control of scale insects, leaf, shoot borers.

Old, senile unproductive mango tree

Unproductive mango orchard

Old, senile, unproductive dense mango orchard

Decline due to dieback

Decline due to lichen infestation

Declined / dried tree

Training to orchard growers

Reiterative pruning of old trees

Application of Bordeaux paste on cut portion

Hard core pruned orchard trees

Orchard floor cleaning

Reiteratively pruned tree

Hard core pruned tree

Sale of pruned wood from rejuvenated tree

New sprouts on trunk

Profuse development of shoots on pruned tip

New shoots

Thinned shoots

Panicle at pruned trunk

Developing canopy on pruned tree

Intercropping in rejuvenated orchard

Developing shoots

Developing shoots

(A) Training for canopy development (A & B) (B)

Flowering in mature mango shoots

Bearing of fruits in rejuvenated tree

Chapter 16

Pomegranate

Pomegranate (*Punica granatum* L.) commonly known as 'Anar', is an ancient favourite table fruit of tropical and subtropical regions of the world. It belongs to family, Punicaceae. There is a great diversity in pomegranate. Originated in southern Asia and said to be native from Iran to the Himalayas in the northern India (Sandhu and Gill, 2013). The orcharding of pomegranate is somewhat expansive and skilful job. The perennial nature of fruit take longer period to come to bearing stage (3-4 years). Thereafter due to neglect in management practices the annual production, many a times compels the orchardist have to go for distress sale of entire poor quality produce. But, the successful commercial fruit production with increased quality production has made significant contribution to economic development of the grower and ultimately those in the growing region in sustainable manner.

PRESENT STATUS

It is found widely cultivated in Mediterranean regions (including Asia, Africa and Europe) since ancient times. The Mediterranean region include. Recently this crop is found to be in extensive cultivation in countries like Spain, Morocco, Egypt, Iran, Afganistan and Baluchistan. It is also grown to some extent in Burma, China, Japan, USA (California) and India. The pomegranate is cultivated as a perennial fruit crop in limited extent in almost all the states of this country, more particularly in Maharashtra, Andhra Pradesh, Uttar Pradesh, Tamil Nadu, Karnataka and Gujarat (Singh, 2007). The versatile adaptability, hardy nature, steady but high yields, better keeping quality, fine table and therapeutic values indicate the avenues for increasing area under pomegranate in India.

Though, India is not a major pomegranate producer globally, but an annual fruit production in this country is about 7,45,000 tonnes only from an area of 1,13,000 ha, the productivity being 6.6 tonnes/ha (ICAR-Data Book-2014), which is quite low as compared to other pomegranate-producing countries (Hiwale, 2009).

PRODUCTION CONSTRAINTS

There is a decline both in quality and quantity of produce after some period of time. As a result of which orcharding becomes economically non-viable and non-remunerative. On the other hand, fruit production over the past few years, has been showing an increasing trend in our country, the productivity is still much below the productive potential with poor quality dominance. Low productivity of this perennial fruit crop has been attributed to various factors, like small size of holdings, preponderance of old and poor productive trees and poor management of inputs such as water, nutrients and pesticides etc. The productivity and quality of this perennial fruit is also continued to remain low due to various problems, the important one is the lack of information regarding skill of pruning and training as well as non-adoption of appropriate agro-techniques leading to large number of existing old and senile orchards turning even unproductive (Sonwane and Desai, 1989; Hare Krishna, 2012). Our country is still technology deficit and what so ever the technologies are available very poor in adoption level. The low production potential that too with declining productivity, attaining age, existing in abundance has become a matter of serious concern for the orchardists, traders as well as scientists (Hiwale, 2009, Hare Krishna, 2012). It has also been observed that loss of tree vigour in older trees result in falling of the foliage and low yields of small fruit, usually accompanied by dieback of twigs and small branches. This may be due to age, disease, soil pests, water induced root damage, and other causes. Moreover old trees harbor pest and diseases, and it is difficult to bring back of the dried and diseased parts into live ones. Uprooting and replacement of this chunk of old orchards will be a long term venture. Majority of farmers have limited access to improved, superior varieties due to lack of knowledge, availability of planting materials and that too for their best cultivation and management practices which further constraint for this fruit crop production. The many existing orchards are established with poor selection of planting material, haphazard plantation and poor management have made many orchards uneconomical (Singh, 2010).

Need of rejuvenation

In pomegranate, its trees witness decline in productivity after certain age making orcharding economically non-viable and non-remunerative. In declining orchard, the productivity status is further affected due to compounding problem of insect-pests and diseases. Those trees/plant are rejuvenated which have attained a stage where they are no more profitable from the grower's point of view (Singh, 2007).

It is essential to take up productivity improvement programmes in the senile plantations, with fresh stock supported with appropriate and integrated combination of inputs, pruning and training techniques. The technology for transformation of old senile orchards for enhanced quality fruit production require well planned and concerted efforts. Though processes and procedures require scientific skill with precision in application/implementation.

The research efforts initiated at various pioneer institutes/ SAUs to standardize a technology for restoring the production potential of existing plantations by a technique called rejuvenation. Rejuvenation needs careful, visionary and scientific approach and it should be followed in skillful manner. The hard core pruning and proper training is a medium term strategy requires scientific skill and technique (Hiwale,2009). Thus existing orchards can be brought back into productive ones through rejuvenation technique, which in turn help in restricting the deterioration of the health of the plant, thereby extending their life and aims at improving the yielding capacity with quality fruits.

REJUVENATION TECHNIQUE

Identification/Selection of Old Orchards

The old orchards become non-productive due to incidence of pest and diseases and neglect are selected. Once the orchard becomes old its productivity goes down, and it is general tendency of the farmers to neglect the orchard. In pomegranate 25% old orchards become unproductive due to loss of vigour in the plants. They need adoption of rejuvenation technology. The cause of decline should be determined and corrected if possible; otherwise, response to rejuvenation pruning will be temporary. If trunk and basic scaffold limbs of trees are not structurally sound due to disease, heart rot, and canker, such rejuvenation procedures are not justified. A limited root system due to hardpan or poor subsoil can result in early tree decline, with trees being more readily stressed by drought. When this occurs, the top becomes out of balance with the more limited root system. Consequently insufficient uptake of water and nutrients by the roots. Hence, selection of old orchards for rejuvenation is also a visionary approach.

Training orchard owners

Creating awareness for increasing adoption of this technology at the initial level, needs proper training. As in pomegranate, rejuvenation can be done by mechanical measures, chemical measures as well as by hard core pruning of plants (Hiwale, 2009). Rejuvenation technology (an invasive technology) should be taught to growers through theory (classroom) and onsite practical demonstration through hard core pruning and the series of processes involved for restoration of youthful vigour, enhancement of quality fruit production, and above all the economic feasibility and period of sustainability. This concern for

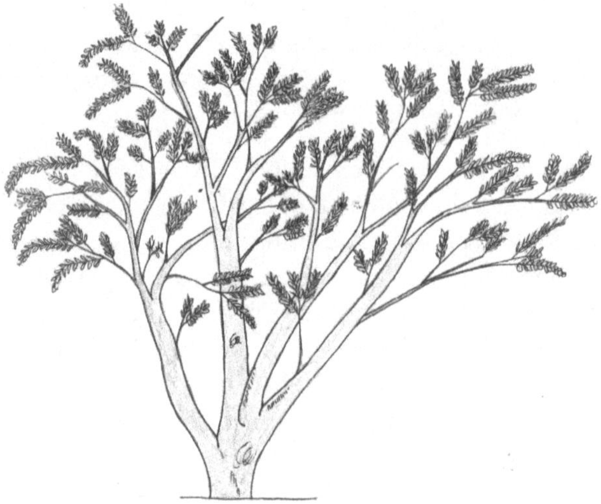

Old senile unproductive pomegranate tree

Pruned tree

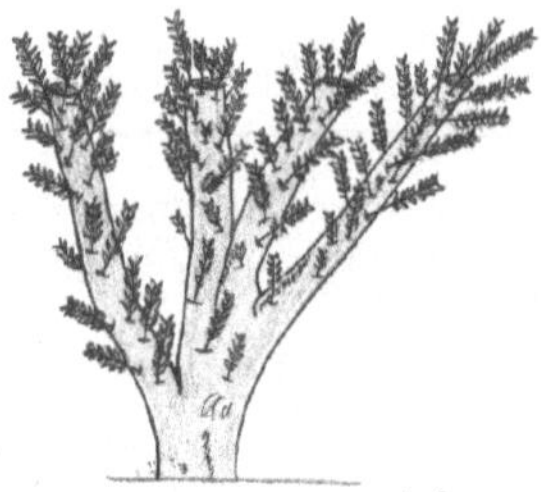

Emergence of shoots

Shoot thinning

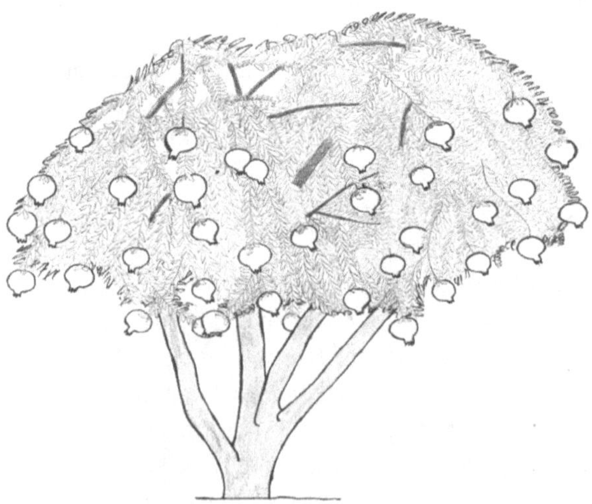

Rejuvenated tree of pomegranate in bearing

the awareness to the growers and enough knowledge base through training. The level of confidence will increase if they are convinced about the economic feasibility, which is vital to increase the level of adoption.

Hard core pruning and training operations

Pruning plays an important role in regulating and controlling growth, flowering and fruiting in perennial fruit trees (Crane *et al.*, 2009; Singh *et al.*, 2012). Hard core pruning for topping off these trees should temporarily restore them to a more favourable top to root ratio. The severity of pruning for rejuvenation depend on the cause and degree of decline. The cutting back the scaffold limbs to height of 0.30 to 1.00 m above the ground, is most severe form of rejuvenation pruning for this crop. It greatly reduces tree size and results in vigorous re-growth. Large wounds remain and production is lost for about a year. It is advisable to paint large cut surfaces on or near the trunk of the tree which result from hedging and topping. Severe topping for rejuvenation which exposes large limbs and trunks that have grown in the shade may result in severe sunburn when done during or shortly before hot, dry weather. In principle, prune perennial fruit trees when they are in dormant phase. In case of rejuvenation, it is important to look for the health, location and value of the fruit trees, which should be determined before going for rejuvenation. Work carried out non-selective pruning at CHES, Vejalpur has shown encouraging results in case of hard core pruning in fifteen year old orchard that was showing signs of decline mainly due to drying of old branches resulting in sparse flowering and fruiting. The plants were headed back to 30cm, 60cm, 90cm and 120cm above ground level in April, no pruning acted as a control. Training adopted as new sprouts per plant was reduced to 8 to 10, by allowing two to three healthy sprouts per branch to balance the framework of the plant. The treatment plants pruned to 30 cm from the ground level showed the best performance. Yield per plant was however has surpassed control in second year itself. Pruning plants to 30 cm from ground level was the best to rejuvenate the old pomegranate orchard and can be recommended as standard practice to rejuvenation old orchards (Hiwale, 2009).

Many mistakes are made when people prune large or heavy branches. In many cases, the pruning process often results in damage to the tree. Pruning of limbs should be careful in order to avoid tearing or ripping of the bark while making the cut. When single cut in a hurry is used, the problems occurs. The final and finishing cut is made at the trunk marked at the desired height, since the weight of the limb/branches has been removed, the final cut can be made with precision and without the risk of damage to the bark of the trunk. Pruning can be done either with manual saw or power operated saw preferably in phase manner from the top. The use of pruning paints and dressings at the cut/wound portion is very important to act as defences against disease and pest infestation. The understanding about the basics of pomegranate trees, when and how to prune is must. The process must be started soon after the emergence of new sprouts, as in this tree, there is a strong desire for a tree-form, the bushy, free-growing shrub develops naturally leading to many unproductive branches. Pomegranate fruits are borne terminally on shoot spurs, arising from matured shoots which have the capacity to bear fruits for 1-2 years. With advance in age, they decline in productivity. Later, a little thinning and pruning of old spurs to encourage growth of new ones is required. Farmers should have always the idea of some useful tips on pomegranate pruning.

Nutrition

Pomegranate productivity is greatly increased by application of manures and fertilizers. Both macro and micro-nutrients affect its growth, development and productivity. After top off pruning for rejuvenation, the half of the recommended adult basal doses of farmyard manure and NPK should be applied after a month gap only. These fertilizer be applied in shallow circular trench of 8-10cm depth around the main tree trunk of half meter radius. Application of about 20kg FYM each tree every year at the onset of monsoon has been found beneficial (Gaur *et al.*, 1971).

Irrigation

After application, the fertilizers are covered with top soil and the plants are irrigated for getting good invigoration effect and early emergence of sprouts. The soil should be well drained. Pomegranate is considered to be highly drought tolerant, its plants respond very well to irrigation. However, water management varies from season to season. Ordinary irrigation practices are quite enough for pomegranate fruit production but drip irrigation be followed to economize water and must be followed in case of rejuvenated orchards. At present, it has become the major irrigation system for pomegranate fruit cultivation in Maharashtra. Nutrient and water management in case of rejuvenated trees are very vital, as it help in restoration of proper growth and vigour for sustainable fruit production in a long way.

Intercropping

During initial 2-3 years after rejuvenation, there is excellent opportunity for economic utilization of inter-space in the rejuvenated orchards as just after the completion of reiterative pruning for rejuvenation, the open interspaces between the plants/trees is created in the orchard like newly planted orchard. Intercrops during summer season and during rabi season have been found most suitable.

Apart from the significant income obtained by the intercrops in rejuvenated orchards, the added advantages like improvement in the soil tilth (health), almost complete check over weed population/growth as well as less incidence of pests-diseases are also received. Water requirements of the intercrops should not clash with those of the main fruit trees. The intercrops selected should not exhaust the soil water and nutrients and should not demand more water than is allowed for rejuvenated fruit trees. The most important aspect is that intercropping should be started at least after one month, when hard core pruning operation is over. The rejuvenated pomegranate trees start commercial bearing after 2 to 3 years and until then intercropping can be done with berseem, lucern, leguminous crops like cowpea, green gram and vegetables like cabbage, cauliflower, beans, peas, tomato, carrot, onion, radish, potato, brinjal etc. However, intercrops should be regulated so that their cultural care does not interfere with the bahar regulation at start of bearing stage. Similarly the intercrop should not help in pest carryover leading to heavy past incidence on the main crop.

Mulching

To avoid occurrence of physiological disorders at after attaining fruit bearing stage, the mulching operation during the summer months is essential, apart from the nutrient and irrigation management, which is mainly to avoid the variation in moisture content of the soil as well. The mulching also take care of the thermal regulation of the rhizosphere region and weed population control (Sandhu and Gill, 2013).

Plant protection measures

Successful pest management programmes developed for this crop should be followed, as the spread of inoculums for infestation and has been attributed to cultural method, regulatory method, chemotherapy and bio-pesticide control. By controlling the insect-pests, the restoration of vigour and keeping the tree more healthy is possible which in turn help in enhanced and sustained quality fruit production (Hiwale, 2009). Intensive care to control the infestation of important pests like bark eating caterpillar, bagworms, leaf eating caterpillars etc., and the diseases like leaf spot, fruit spot, fruit rot etc. Control methods may be biological, mechanical or chemical or the combination depending upon the growers preference and specific circumstances. Insects start damaging right from the beginning of the rejuvenated trees, hence care is required from that moment itself, as these are attracted to new vegetative growths *i.e.* leaf flushes from May to August and the spray schedule of Monocrotophos @ 2ml/l or Cypermethrin @ 1ml/l 2-3 times during new flush emergence *i.e.* during the month of July-August can save the damage from these insects. The serious damage is being observed by the attack of bark eating caterpillar, which tunnel into the branches and trunk causing decline and even death of the tree, can be controlled by the spray of insecticide at initial stage. But at tunneling stage, it is require to clean the fresh frass and webbing, finding the hole, notching the hole with hard pointed spoke/poker, plugging with DDVP/Nuvan (0.1%)/Kerosene oil soaked cotton and ultimately sealing the hole with mud. Spraying of copper fungicide during the month of May-June, August-October and March-April takes care of most leaf spot and other algal/fungal diseases. Brushing/pasting of Bordeaux paste to the stump of the main trunks/branches from the bottom keep the plant clean and saves the crop from many pests infestation. Perhaps the most serious problem with pomegranate is the occurrence of a fungal disease which affects both the leaves and the fruit, causing premature leaf loss and also resulting in fruit splitting on the plant. An application of white latex paint and fungicides using a power sprayer is more efficient where large numbers of trees are involved. While the leaf drop may be ignored, fruit splitting cannot, since the splitting usually occurs just as the fruit begins to mature. However, copper fungicide use during fruit development of late spring through summer may alleviate the problem of leaf drop and fruit quality.

Management of Physiological Disorders

Sun Scald

Surface skin of fruits facing afternoon sun turns brownish black due to scorching, underneath skin is normal. It is required to develop proper canopy architecture to avoid the direct exposure of fruits to sunlight. Spraying Kaolin during the hot summer months is useful in reducing the sunscald. First spray of 5% and subsequent 1 or 2 additional spray with Kaolin @2.5% at 15 days interval reduce the sunscald. Shading 35% is effective in reducing sunburn on pomegranate fruit. Bagging the fruits with butter paper covers is useful in minimizing fruit spoilage due to sun sunscald. White colour bags are more effective in reflecting sunlight and protecting the growing fruits.

Internal Breakdown of Arils

Randomly open some of the fruits and observe soft arils of brown colour in a fruit looking healthy from outside, more common in dry condition with intense heat. The harvesting should not be delayed beyond maturity period. The pomegranate fruits should be harvested as soon as they mature.

Fruit cracking

Healthy fruits split open even though they do not have any scar/spot on the fruit surface , generally when rains come or irrigation is given after a long dry spell.

Apply adequate and regular irrigation during fruiting season. Sprays of gibberellic acid (GA_3) @120 ppm, benzylaminopurine (BA6) @ 40ppm, boron (0.2%) at fruit enlargement and 1 month before harvest help in reducing cracking. Cover fruits with butter paper bags in hot dry period. Maintain adequate calcium and potassium levels as per soil test values.

Little leaf

The frequent use of this weedicide glyphosate at high doses (>4ml/l) results in reduction of leaf lamina and internodes, which leads to extreme reduction in leaf size forming little leaves, rossetting of leaves and stunting of the entire plant. The symptoms resemble mycoplasma or viral diseases. Use weedicide judiciously. Spray when no wind is there to avoid wind drift.

Pest Management

Pomegranate Butterfly/Fruit borer (*Deudorix isocrates*)

It is the most important pest and distributed all over India, wherever pomegranate is grown. The adult lays eggs at the time of fruit setting in the flowers. Larvae develop inside the fruit and these larvae bore out of the fruits. Black excreta of caterpillars are seen commonly coming out of exit holes or stuck around the holes. The affected fruit is of no use and may rot or fall down.

Remove infested fruits and destroy. Spray crop with Monocrotophos (0.1%)or Phosphamidon (0.03%) or Carbaryl (0.4%). Repeat the sprays with Fenvalerate (0.01%). First spray should be given at the time of fruit setting.

Mealy bugs (*Planococcus* sp., *Ferrisia* sp.)

Adult bugs and their nymphs are seen on all parts. They suck sap from leaves, twigs, flowers and fruits. The insects form thick encrustations on fruit which deteriorates fruit quality. Under severe attack flowers and fruits fall down. Cottony appearance of Mealy Bug and sooty moulds are the visible symptoms of mealy bug infestation.

Thrips (*Rhiphiphorothrips cruentatus*)

Thrips tear leaves, flower stalks, petals, sepals and suck the sap that oozes out, resulting into leaf curl, drying of young shoots and shedding of flowers. On fruits discoloured areas are seen which later show rusty scars.

Mites (*Oligonynchus punicae, Aceria granati*)

Shiny white/brown patches on undersurface of leaf are due to mite infestation. Adult and nymphs feed on the lower leaf surface. The affected leaves may curl and dry.

Scale insects (*Parasaissetia nigra*)

They are found attached to all plant parts and weaken the plant resulting in flower and fruit drop.

Aphids (*Aphis punicae*)

It is a serious problem on new flush. Aphids infest tender shoots, flowers, leaves, flower buds and fruits. Severely infested leaves curl. They suck cell sap. The affected parts get discoloured and disfigured. These insects secrete copious amount of honey dew, on which sooty mould develops, hindering photosynthetic activity of the plant.

White flies (*Siphoninus phillyreae*)

They severely infest leaves on the dorsal sides, resulting into curling, yellowing and drying.

Affected plant parts should be removed and destroyed on observing the insect infestation. Spray monocrotophos (0.1%), cloropyriphos (0.02%), immidacloprid (0.04%), deltamethrin (0.1%) and dichlorovos (0.05%) as prophylactic or on observing the symptoms.

Bark-eating caterpillar (*Inderbela quadrinotata, I. tetraonis*)

The caterpillar bores the bark and feeds under webbed galleries of silk and excreta. Old trees or those trees which are not properly maintained are usually attacked. Usually one caterpillar in each gallery is seen, ther may be 10-12 galleries in a badly infected tree. Such trees bear no fruit and dry up.

Orchard should be kept clean and overcrowding of trees should be avoided. Whenever larvae holes are observed, inject larvae holes with quinaliphos (0.01%) or fenvalerate (0.05%). Give a prophylactic spray or on observing the symptoms spray with carbaryl (0.04%) or dichlorovos (0.08%) or monocrotophos (0.1%).

Stem-borer (*Celosterna spinator*)

The beetle is found to infect main branches or trunk in a neglected orchard. It makes a hole and bores through the bark. Holes are seen on the bark and network of dried excreta is seen outside the hole. It comes out at night and is often seen during rainy season.

Treat the holes produced by borer with dichlorovos (0.25%) or monocrotophos (0.25%) and seal holes with clay. Spray quinalphos (0.05%) or chlorpyriphos (0.05%).

Shoot hole-borer (*Xyleborus formicatus*)

Beetles bore holes in roots and trunks resulting in drying of trees. Their incidence increases in rainy season. When insect symptoms are observed (yellowing and drying of plants) soil drenching with chlorpyriphos (0.02%) or monocrotophos (0.15%) in and around the affected trees is effective. As prophylactic, spray monocrotophos (0.1%)or chlorpyriphos (0.2%) or carbaryl (0.2%).

Leaf footed plant bug (*Leptoglossus* sp.)

Adult bugs pierce the fruits with their long proboscis and eventually the fruit dries and falls down.

Harvest fruits before ripening. Destroy fruits fallen on ground regularly. Spray crop with monocrotophos (0.1%) or phosphamidon (0.03%) or carbaryl (0.4%), repeat the sprays with fenvalerate (0.01%).

Leaf-eating caterpillar (*Achaea janata*)

Freshly hatched larvae damages the leaves by voraciously feeding on leaf lamina and adults suck juice from the fruits.

Spray crop with monocrotophos (0.1%) or chlorpyriphos (0.1%) as prophylactic spray or on initiation of insect infestation.

Fruit flies (*Batrocera zonata*)

On pomegranate these flies are of minor importance. They feed on vast variety of hosts. They lay eggs just below the epidemis of ripening fruits. On hatching the small white grubs feed inside the fruits, which rot and fall down. Harvest fruits before they ripen as the flies are not able to puncture the hard rind of unripe fruits.

Harvest fruits before ripening. Destroy fruits fallen on ground regularly. Rake soil between trees. Spray crop with monocrotophos (0.01%)or phosphamidon (0.03%) or carbaryl (0.4%), repeat the sprays with fenvalerate (0.01%).

Fruit sucking moths (*Othreis ancilla, O. cajecta, O. fullonica, O. maternal, Achaea janata*)

It punctures mature fruits during dusk by piercing mouth parts and sucks juice. Secondary infections through these punctures result in fruit rot. They lay eggs on weeds, therefore keep orchards weed free.

Put poison bait containing malathion 1 ml + 100g jaggery+ 5 ml vinegar in small tins during night. Generate smoke in orchards during dusk time.

Termites-white ants (*Odontotermis obesus*)

These feed on many crops. They feed on all things which contain cellulose. It is more serious in light or light loamy soils and dry areas where proper facilities for irrigation are not usually available. They do not survive under anaerobic conditions. It is more serious in Rabi season. It feeds on roots resulting in drying of leaves, withering and drying of whole plant.

Use well decomposed organic manure. Remove dead decaying matter or dry stubbles from the field to avoid termite infestation. Flood irrigation is useful in protecting crop from termite. Treat soil with quinalphos 1.5% or methyl parathion 2% dust @ 25 kg/ha before planting. In standing crop apply chlorpyriphos 20 EC @ 4 L/ha with irrigation water for effective protection of crop against termiteinfestation.

Nematodes (*Meloidogyne incognita, M. javonica*)

Almost 12 genera of nematodes are known to be present in pomegranate. Meloidogyne incognita and M. javonica are most widespread and prevalent. The roots of infected plants form knots of very small size. They result in weakening of plants. Apply phorate (@40 gm/ Plant), or carbofuran 2 kg a.i./ha.

Management

Bacterial Blight

On leaves water soaked spots on undersurface of leaf which, later turn blackish brown irregular spots of different size on leaves, with a clear regular chlorotic halo against light.

On fruits spots of water soaked appearance or brownish black spots with cracks or split fruits.

The bacterial ooze comes out after rains/spray/dew, which feels sticky to hands and after drying gives white shiny encrustation on the surface of blight lesion. On stems water soaked grey/black lesions on nodes or brown cankers with/ without stem cracking.

Orchards where blight has been detected or present in severe proportions or orchards which are disease free but in blight prone areas should follow the schedule strictly. Avoid mrig bahar (rainy season) crop and shift to hasta bahar crop for at least 3-4 few years.

Follow all sanitation measures. Remove fallen plant debris and burn them-do not dump them in or near orchards nor throw them in irrigation channels. Drench bleaching powder (a.i. 33% cl) every 3 months @25kg/1000 litre water/ha on ground below the canopy. Disinfect pruning tools- secateurs etc. after handling each plant with sodium hypochlorite (2.5%) and keep orchard free from weeds. Practice proper pruning and training.

Apply Bordeaux paste (10%) to the cut ends after pruning. Oil based pastes [COC paint made by mixing 500g COC + 1L linseed oil or Chaubatia paste prepared by mixing 1 kg red lead (non setting grade)+ 1 kg copper carbonate + 1.25L linseed oil] are preffered for pasting during rainy seasons.

Wilt

Plants with yellowing / drooping/ drying of leaves in some of the branches or entire plant. Observe the roots and split open the roots and lower portion of the stems, if you observe brown/gray/black discolouration of wood. It is only xylem is brown (*Fusarium* sp.). Pin holes are observed it is shot hole borer. Knots are observed on fine roots it is nematode infestation.

On observing first symptoms of wilt first ascertain the causes.If it due to fungal pathogens in the orchard immediately drench soil with chlorpyriphos 20EC (2.5ml/ 1 TO 4.0ml/l) + carbendazim 50WP (2.0g/l) or propiconazole 25EC (2ml/l) use 5-8 l solution/plant. Also drench at least 2-3 healthy plants on all the four sides around theinfected plants, repeat the drenching 3-4 times at 20 days interval. Drenching with metalaxyl or dithane M-45 (2g/l) will be beneficial if *Phytophthora* is causing any loss.

For controlling shot hole borer (Xyleborus spp.) which is associated with wilt disease, 10 litres preparation containing red soil (4kg) + chlorpyriphos 20EC (20ml/l) +copper oxychloride (25g) needs to be applied on plant base up to 2 ft. from second year onwards. To control stem borer, inject in the holes on the trunk with DDVP 2-3 ml and plug the holes with mud.

Wilt due to root knot nematodes can be managed with soil application of phorate 10 G@ 10-20 g/plant or carbofuran 3G @ 20-40g/plant in the plant basin, in a ring near root zone and cover it with soil. Drenching with azadirachtin (1%)@ 2ml/l is also recommended. Plant *Tagetes erecta* (African marigold) between plant to plant space in a row, or in a ring, on the border of plant basin. For effective results these should be grown for more than 4-5 month.

Fungal Spots and Rots

Cercospora Fruit spots : Dark black discrete spots of various sizes without cracks and no stickiness.

Fruit Scab *:* Rough raised brown spots giving russet appearance to fruit skin.

Anthracnose **(on Fruit)** : Hard dark black to brownish black spots of no proper shape or size, sometimes spots with light centre and dark brown black edges.

Colletotrichum rot : Fruits start discolouration from calyx end or fruit surface. The discoloured area becomes reddish brown to black and is dry. The infected area covers half to full fruit within a week. The rot extends beyond rind into the arils which disintegrate and are dark grey/ brown black colour but not watery.

Alternaria heart rot : There are no external signs of heart rot, however, due to disintegration of arils the fruits may appear softer on application of slight pressure and in later stages become light weight. Rot is revealed only when the fruit is opened. The arils inside the fruit turnbrown-black. Greyish black fungal growth/ sporulation may be observed in completely rotten fruits on the rotten arils. The rind is generally not attacked. Slight rind discolouration may be seen in severe cases or when secondary rot fungi are associated.

Aspergillus rot : The pathogens results in brownish soft rots, with a depressed centre in later stages where blackish sporulation of the fungus can be seen. These generally may grow inside without any external symptoms for some time. The rind and underlying pulp became soft, but fruits retain their shape.

Penicillium rot : This pathogen generally starts on the wounds and produces light tan coloured mushy soft rots, with bluish green mould growth in later stages.

Phomopsis rot : Dark blackish brown dry hard rot with distinct sharp edge, generally starting from calyx end and spreading towards stem end of fruit. When flowers and young fruits are affected they drop. On bigger fruits yellowish brown toblack spots appear all over the fruit.

Phytophthora blight/rot : The pathogen Phytophthora spp. Affects seedlings, foliage as well as fruits particularly during rainy season, when humidity is high. The affected seedlings show damping off and wilt symptoms, whereas, leaves and young leaves and young twigs give typical blighted appearance. The fungus also attacks flowers and fruits at all stages and also causes fruit rot. Light brown tan colour rot later turning darker, which is neither hard nor very soft, generally appears frist in fruits near ground level and spreads rapidly with white sporulation in humid conditions.

In orchards, where fungal spots, rots or *Phytophthora* blight make yearly appearance and adversely affect fruit quality or marketable yield, change flower regulation time from rainy to winter season.

Maintain orchard sanitation by removing plant debris, pruning of dead dry branches, drenching of ground with 2.5% bleaching powder (33.3%Cl) @ 25 kg/1000 litre water/ha and sterilization of pruning tools with 2.5% sodium hydroxide.

Sprays with suitable chemicals at recommended doses, starting from flowering at 07-14 days depending up on season and nature (systemic/ non systemic) of fungicides give good control.

During rest period Bordeaux mixture (1%) alternated with copper or other economical fungicides at 15 days period helps reduce the inoculum load of pathogens for next crop season.

Leaf and fruit spots as well as rots caused by different fungal pathogens are effectively managed by the sprays of carbendazim (0.2%), propiconazole 0.1% thiophanate methyl (0.15%), mancozeb (0.2%), difenconzole 25 EC (0.05%), captan (0.2%), benomyl (0.2%), ziram (0.25%), Bordeaux mixture (0.5-1%) and copper oxychloride (0.25%).Combination treatments of carbendazim (0.1%)+ mancozeb (0.2%) and benomyl (0.1%)+ mancozeb (0.2) also improve fruit quality and yield.

Anthracnose and calyx end rot of pomegranate caused by *C. gloeosporioides*) can be effectively checked with Tricyclazole 18% + Mancozeb 62% WP @ 2.5g/l. Fruit dip with dithane Z-78 (0.1%), captan (0.2%), benomyl (0.2%) reduce post harvest rots.

For Phytophthora blight/rot, sprays of metalaxyl 8% + mancozeb 64% (0.25%) or mancozeb (0.25%) or dimethomorph (0.1%) as soon as disease appears or as prophylactic in orchards where it makes yearly appearance are effective, Sprays of Fosetyl Al 80% WP (0.2%), formulation containing cymoxanil 8% + mancozeb 64% (0.2%) or dimethomorph 9% WP + mancozeb 60% (0.15%) also give effective control of Phytophthora.

Fruit Yield

Fruit yield and physicochemical characteristics of mature fruits were found to be better in fruits obtained from rejuvenated trees came into bearing. Maturity period is found to be slightly delayed in rejuvenated plants. The tree starts yielding from second year onwards giving 70-75 fruits per tree. Yield increases progressively from 3rd to 5th years and a tree may produce on an average 175-200 fruits per year. The average yield in a further well managed orchard may be as much as 200-250 fruits per tree. Economic yield obtained continuously for another 10-15 years after rejuvenation.

Economics

The various operations in rejuvenating old senile trees is not an immediate process, but usually it can be done effectively over a period of 2-3 years and including maintenance and management aspects. The cost of hard core pruning and subsequent operations in one hectare of old senile orchard have been estimated (Table-1). It has been found that the old senile orchard can be rejuvenated back just like young bearing orchard within a time gap of 2-3 years without any income loss in-between by adopting and following the technology. Economics of rejuvenation of old pomegranate orchard have been clearly worked out in 3-4 consecutive years with the set particulars of operations like sale of fruit before rejuvenation, cost of hard core pruning, sale of woods, cost of intercropping, profits from intercrops, sale of fruits from rejuvenated trees and finally gross income have been found very encouraging with high benefit cost ratio with other added advantages. Economics of rejuvenation (tentative) of old pomegranate orchard is given in Table-1.

Table 1. Economics of rejuvenation of old pomegranate orchard (one hectare).

Operation	Year I	Year II	Year III	Year IV
Sale of fruits before rejuvenation	25000	-	-	-
Cost of hard core pruning	10,000	-	-	-
Cost of orchard floor cleaning	4,000			
Cost of inputs and application charges	10,000			
Sale of woods	10,000	-	-	-
Cost of intercropping	25,000	30,000	20,000	20,000
Profit from intercrops	50,000	60,000	35,000	35,000
Sale of fruits from rejuvenated trees	-		80,000	1,60,000
Gross income	**36,000**	**30,000**	**95,000**	**1,75,000**

Impact of Technology

Hence, with the scientific skill and approach, complete transformation can be brought about by changing the exhausted phase of the old trees to behave and bear like young commercial trees with sustained performance. This transformation gave great impetus and boost to the enhanced quality fruit production having high economic returns. This wonder became possible only with the adoption of scientific rejuvenation technology. It is indeed became an example, when the utmost transformation tried through rejuvenation pruning of old senile, unproductive orchard, without destroying the actual identity and fortunately, only in time gap of two to three years it started behaving like young commercial orchard with sustained economic returns. As a result, the increasing trend of production and productivity of quality yield in coming years as commercial bearing orchards, the fruit growers having old plantations giving uneconomical yield convincingly started adopting the technology developed by various Research Institutes/State Agricultural Universities. The success of rejuvenation technology in transforming the old orchards into young commercial bearing orchards have become the most important consideration in National Policies of Government Horticulture Developmental Plan.

Old, unproductive pomegranate trees

Hard core pruning for rejuvenation

Rejuvenated tree in bearing

Quality fruits on rejuvenated tree

Chapter 17

Sapota

Sapota or sapidilla or bully tree or *chiku* (*Manilkara achras* (Mill.) Fosberg) is a delicious tropical fruit crop, was introduced from tropical America. Among lesser-known fruits, area expansion and production of sapota have been observed spectacular. During as the area under sapota is fast increasing and this crop is attaining the status of major fruit crop.

PRESENT STATUS

India is considered to be the largest producer of sapota in the world, though it is considered to be a minor crop in India (Bal, 1997). The area under sapota in India during 2000-01 was 69,400 ha with a production of 6.74 lakh tonnes, while during the year 2012-13, it was 164000 ha with a production of 14.95 lakh tonnes (ICAR-DataBook 2014). The area under sapota is increasing so rapidly owing to a wider range of adaptability. In many regions, cultivation of sapota is preferred over mango (Bal, 1997). The important sapota growing states are Maharastra, Gujarat, Andhra Pradesh, Karnataka, Tamil Nadu, Kerala, Uttar Pradesh, West Bengal, Punjab and Haryana. Recently, our country has made a beginning in export to Middle East countries.

PRODUCTION CONSTRAINTS

The long pre-bearing phase, the large stature of the trees, poor fruit set, flower and fruit shedding are some of the production problems in sapota which are required to be managed through proper canopy management and crop regulation practices. The adoption of proper training and pruning of trees, use of bio-regulators, irrigation and nutrient management need attention for commercial fruit production. Though sapota is considered as a hardy crop which requires less attention, the national average productivity is very less (14.19 tonnes/ha). Among production problems, the occurrence of certain physiological disorders, apart from the problems of pests and diseases, which reduce the quality of the fruit drastically. Till date, it is meagrely exploited at R&D level and production mainly constrained due to narrow genetic base, lack of technology for rapid multiplication, long pre-bearing phase, poor fruit set, lack of technologies for canopy management, lack of concepts and related technologies for rejuvenation of old declining orchards, problem in harvesting and storage and pests and diseases management. All these problems have already received the attention of the scientists and suitable management technologies are being evolved through field trials at different research centres.

Need of rejuvenation

Sapota is an evergreen tree and normally grows up to a height of 20 m. The different cultivars of sapota with varied types of branching habit grow old. The spreading canopy starts bearing fruits of inferior quality. Because of the unmanageable size of the trees and the different growing habits, the canopy management through pruning and training is very essential in sapota to increase the productivity. As the age advances, mere neglect of the orchards become unproductive, with the start of declining phase. The rejuvenation technology is very much required for making the productive orchard with sustainable high quality fruit production.

TECHNIQUE OF REJUVENATION

Identification/Selection of Old Orchards

In old unproductive orchards, the flowering and fruiting occur negligibly and that too of poor quality, suffers due to the attack a large number of pests, diseases and disorders. The decline in productivity of the tree also increases in unmanaged orchards with advancing age.

Training to orchard growers

In sapota, flowers are borne on past season shoot. The flowers are polygamous in nature and hence planting of single sapota tree is not recommended. Annual pruning in commercial bearing orchard and hard core pruning for rejuvenation of old senile unproductive orchard trees require skilful and timely operations. There must be clearly spelt to growers to enjoy the potential bearing of quality fruit production. The information and aftercare including thinning of shoots, training of branches for canopy re-building at the initial two years along with proper input application and management practices are required for rejuvenation.

Pruning and training

No definite system of training has been adopted by the growers of sapota. But it is required to train the trees in central leader system. The reiterative pruning for rejuvenation, plants are topped to 1.00 to 1.50 m above the ground level.

After emergence of new shoots below the cut point, 3-4 well spaced scaffold limbs are selected and allowed to grow to make a strong framework. The lower most shoots should be removed properly up to skirting height. At times, thinning of branches is affected in old plantation. If pruned trees are trained to a well balanced distribution of branches gives well shaped dense canopy. Though thinning of crowded branches should be done skilfully. Pruning in sapota is confined to open trees to light and remove dead and diseased branches. The same trend is being maintained after hard core pruning in old senile unproductive orchards.

Nutrition

After the operation of reiterative pruning in phased manner, application of manures and fertilizers as commercial bearing trees are to be applied in half quantity after a gap of at least 25-30 days and the remaining half after 6 months. In Gujarat, Maharashtra and Karnataka, manures and fertilizers are applied mainly in two splits. Many recommendations have been made as under rainfed condition, it is advised that sapota trees should be fed at the rate of 1.5 kg N/ tree/year in the form of castor cake or urea. The application of 600g N and 400g K /plant in PKM 1 sapota recorded higher yield of 29.50 kg/tree. In another trial, the treatment with 5 kg vermicompost + 200:40:150g NPK/tree recorded maximum tree height of 4.08m and maximum fruits yield of 60.93 kg/ tree.

Irrigation

Irrigation should be a regular feature at the initial year. The pruned trees should be watered regularly during dry season to maintain enough moisture content in the rhizosphere and at an interval of 6-12 days during winter. In general excess irrigation may cause flower and fruit shedding increasing the number of mis-shapen fruits. Drip irrigation at the rate 20-30 l/day/tree is recommended by Tamil Nadu Agricultural University, Coimbatore. It should be followed with lower dose only 20-25 days after the pruning, *i.e.* one-third of the recommended dose of water may be given and slowly, the quantity should be increased as the vegetative growth advances. Fertigation studies with water soluble fertilizer indicated that fertilization with 80% of water soluble fertilizers at bimonthly intervals yielded 5800kg/ha as compared to 4300 kg with 100% normal fertilizers applied through soil. Fertigation economized the use of fertilizers and water to the extend of 20-25 percentage (Shivashankar and Khan, 1994).

Mulching

Sapota grows well in warm and moist climate. Weeds compete for nutrients and moisture hence should be removed frequently and can be kept under control under the tree canopy by mulching.

Intercropping

Intercrops like banana, papaya and plum may be planted as filler crops in between the rows in those orchards which has been reiteratively pruned for rejuvenation. The vegetable crops like okra, brinjal, tomato, chilli, onion and leguminous crops like cowpea, green gram, black gram and cluster bean can be grown as intercrop in sapota. It also controls the weeds and improves soil tilth due to intercultural management.

Managing Physiological Disorders

Fruit drop

Sapota has the problem of low fruit setting and shedding. Only 10-12% of the total fruits set, develop and retained until maturity. Most of the fruit-drop occurs immediately after fruit setting. Increase in fruit setting and retention are possible by spraying NAA and GA_3 at 25-100 ppm during flowering and at 15-day interval (Rathod and Amin, 1981).

Corky tissue

Corky tissue (CT) is one such disorder in Cricket Ball variety of sapota, which hamper yield upto 35% in certain seasons. Recently, the increasing global warming, leading to increase in temperature of earth's atmosphere and variations in the rainfall pattern have affected the integrated development of plant and eventually appear as disorders in fruit crops. Corky tissue (CT) of sapota is a physiological disorder characterized by hard lump in the pulp, slightly desiccated in nature and acidic to taste. This disorder shows no distinct external symptoms and becomes visible only when the fruit is cut open. Under extremely severe conditions, corky skin eruptions are seen (Sulladamath, 2005). As such, the edible quality of the affected fruit is poor and in advanced stages, it becomes poor for human consumption.

Flat limb / fasciation

This disease is caused by *Botryodiplodia theobromae* Pat. It has become prevalent in almost all sapota-growing states but etiology of fungus remains to be worked out, as consistent reports are still lacking. However, affected branches become flat and twisted with rough marking on affected tissues. The leaves become thin, small and yellow and there is clustering of leaves and flowers at the top of affected twigs. Flowers remain infertile and fail to set fruits and even if fruits are set they remain small, hard and unripe. In severe case, leaves fall prematurely giving barren look to flattered branches and sometimes the thin shoot appears from the tip but finally dries out. It is recommended to prune and destroy the affected branches for controlling the disease.

Phanerogamic parasite

Phanerogamic parasite (*Loranthus* sp.) is most common in unattended sapota orchards and sometimes become the major cause of tree decline. At first, it reduces yield by reducing fruiting area and fruits also become of poor quality. The parasite is a glabrous shrub with green foliage but without true roots. Stem of the parasite is thick and erect, and arise in cluster at the point of attack. This can effectively be controlled by cutting of the affected branches with tumors and their destruction.

Management of Pest

Stem-borer / bark-eating caterpillar (*Indarbela tetraonis* Moore)

The infestation of fruit-borer can immediately kill the rejuvenated trees, if not cared properly. The grub bores into the bark of sapota trunk and feeds on living tissue inside the bark. The presence of borer of insect can be detected from the chewed bark as woody frass thrown out of a hole in the trunk.

Kill the insect by thrusting a stiff wire into the tunnel. Plug the hole with a wad of/inserted cotton wool soaked /dipped in kerosene oil or Nuvan solution and seal the hole with mud, so that due to suffocation the borer dies inside.

Fruit fly (*Bactrocera (Dacus) dorsalis*)

The damage caused by fruit fly as its yellowish apodous maggot stage bore into semi-ripen fruits with decayed spots and dropping of fruits, cause oozing of fluid with brownish rotten patches on fruits. It causes significant yield loss by spoiling the quality of fruit.

Collect fallen infested fruits and dispose them by dumping in a pit and covering with soil. Provide summer ploughing to expose the pupa. Monitor the activity of flies with methyl eugenol sex lure traps. Use bait spray combing any one of the insecticides and molasses or jaggery 10g/litre, Fenthion 100EC 1ml/litre, Dimethoate 30 EC 1 ml/litre, two rounds at 2 weeks interval before ripening of fruits. Prepare bait with methyl eugenol 1% solution mixed with carbaryl 50 WP 4 g/litre. Take 10 ml of this mixture per trap and keep them in 25 different places in one hectare. In heavy infestation - application of dust and sprays of Pyrethrum is advised. Spray Fenthion 100EC 2ml/litre. Release natural enemies (*Opius compensates* and *Spalangia Philippines*) in field.

Hairy caterpillar (*Metanastria hytaca*)

The damage caused by the larva stage of this pest, which is yellowish brown with black spots and long lateral tufts of hairs caterpillars, which feed on leaves causing defoliation.

There is need to spray Phosphamidon 40 SL 2 ml/litre. Application of neem oil 2% or neem seed kernel extract 5% as spray is also advised.

Stem borer (*Plocaederus ferrugineus*) : The adult stage of its medium sized reddish brown beetle cause damage. The symptom of damage is visualized by the presence of small holes at the collar region, showing symptoms ofgummosis as well as extrusion of frass through the bore holes at the collar region. Infestation also cause yellowing and shedding of leaves and drying up of twigs and gradual death of the tree.

There is need to adopt field sanitation. Swab Coal tar + Kerosene @ 1:2 or Carbaryl 50 WP 20 g/litre on the basal portion of the trunk (3 feet height). After scraping the loose bark to prevent oviposition by adult beetles. Padding with Monocrotophos 36 WSC 10 ml in 2.5 cm /tree soaked in absorbent cotton. Apply Carbofuran 3G 5 g/hole and plug with mud.

Scale insect (*Pulvinaria psidii* Mask) : The scale insect suck the sap by infesting along the sides or mid rib and surface of leaves and twigs. The scales are green or brown in colour and oval shaped.

Spray Chlorpyriphos 20EC or Phosalone 2 ml/litre or Cypermethrin 0.025%.

Leaf miner (*Anarisia gemoniella*)

The tiny caterpillar of a green or greyish moth mines into the surface of young leaves. Affected leaves curl up, irregular galleries or mines are seen on the surface of leaves and sometimes caterpillars are found inside the mines or curled leaves. Later on, affected leaves get destroyed, dry up and fall.

Sapota leaf webber (*Nephopteryyx eugraphella*)

This is an important pest on sapota. The larvae constructs a tunnel of webs and grass and feeds on the green matter of leaves. All plant parts like leaf buds, flower buds, tender leaves and fruits are attacked. Occasionally three or four larvae may be seen in a tunnel.

Field sanitation is quite essential. Keep orchards free from weeds and debris. Collect and destroy the egg mass regularly. Burning the groups of larvae found on tree trunks with torches. Spray Chlorpyriphos 20EC or phosalone 2 ml/litre or Cypermethrin 0.025%. Dusting Carbaryl 10 D on the trunk and branches (around the tree 4 feet) helps control. Field release of Chalcidid was p, *Brachymeria* sp., is helpful.

Management of Diseases

Sooty mould (*Capnodium* sp.)

The casual fungus disease develops as the honey dew like excretion on the surface of the leaves. This ectoparasitic disease adversely affects the photosynthesis function of leaves and disfigures the fruit (Rangaswami, 1984).

Phavophleospora leaf spot (*Phavophleospora indica*) : The infestation starts as yellow coloured dots few mm in diameter, which gradually increase in size and are surrounded by violet coloured ring which later on becomes brown coloured.

Botryodiplodia rot (*Botryodiplodia theobromae* Pat.)

The infestation starts as water soaked spot at the stem end. It enlarges gradually and extends towards the styler end and within 8-10 days, the fruits become completely rotten. This fungus has been found responsible for causing leaf spot in sapota als. This disease is characterized by the production of soft watery rot, completely rotten fruits become shapeless.

Phomopsis soft rot (*Phomopsis sapotae*)

This fungus also causes leaf spot and soft rot on fruits and the infestation found to occur mainly from the months of November to March. The infection starts from the stalk end and the disease develops as a water soaked spot which extends and covers the whole fruit causing soft rot.

Pre-harvest spays of Bavistin @ 2g/litre and/or treating the plucked fruits with 1000 ppm Bavistin solution dip controls the disease.

Fruit yield

After the second year of rejuvenation, cultivars like Cricket ball, Calcutta Round start bearing fruits with larger sizes. The average yield start, after fourth and fifth years as 10-15 tonnes/ha. The lowest tree height is obtained with medium pruning in sapota PKM 1 results in better quality fruits. Sapota is a climacteric fruit and it improves in quality after harvesting but immature fruits should never be harvested.

Old & unproductive tree

Dense branching / stems

Sprouts on pruned stems

Hard core pruning

Unproductive dense branching tree

Damage due to dense branching

Well developed canopy in rejuvenated trees

Fruiting in 2nd year after pruning

Fruiting in 3rd years after pruning

Flowering on rejuvenated tree

Fruiting on rejuvenated tree

References

Abrahamson 1979. Patterns of resource allocation in wildflower populations of fields and woods. *Am. J. Bot.* **66**: 71-79.

Alam Syed Nurul 2011. Major Insect Pests of Jack. *In: The Jackfruit* (Ed Sudha George Valavi, K V Peter and George Thottappilly). Studium Press LLC.USA. pp.191-196.

Alila Pauline. 2012. Mango in eastern and northeastern states of India. *In: Mango (Vol. 2) cultivation in different countries,* pp 28-41, Sudha G Valavi, K Rajamohan, J N Govil, K V Peter and George Thottappilly (Eds.), CABI Publications, UK.

Amma S Prasanna Kumari and Kumaran K 2011. Jackfruit in South India. *In: The Jackfruit* (Ed Sudha George Valavi, K V Peter and George Thottappilly). Studium Press LLC.USA. pp. 393-408.

Anila R and Radha T. 2003. Physicochemcal analysis of mango varieties under Kerala condition, *Journal of Tropical Agriculture* **41**:20-2.

Anila R and Radha T. 2003. Studies on fruit drop in mango varieties. *Journal of Tropical Agriculture* **41**: 30–32.

Ansari A M, Akhilesh S and Ekhkaque A. 2011. Studies on rejuvenation of poor bearing citrus plants. *Progressive Agriculture* **11**(2):491-4.

AOAC. 1990. Official and Tentative Methods of Analysis. Association of Officinal Agricultural Chemists (15th Edn.), Washington, DC, USA.

Ashmun, J. W., and L. E Pitelka. 1984. Light-induced variation in the growth and dynamics of transplanted ramets of the understory herb, Aster acuminatus. *Oecologia,* **64**: 255-262.

Awasthi, A.M and Mishra, R. (1969). Effect of pruning on subsequent vegetation growth, fruit set, fruit drop and quality in ber (*Zizyphus mauritiana Lamk.*) *cultivars. Punjab Hort. J.*, **9** (1-2):54-60.

Baba J A, Akbar P I and Vijai K. 2011. Rejuvenation of old and senile orchards : A review. *Annals of Horticulture* **4**(1):37-44.

Bakhshi J C, Uppal D K and Khajuria H N. 1988.The pruning of fruit trees and vines. Kalyani Publishers, pp. 1-153.

Bakshi, J.C., Uppal, D.K. and Khajuria, H.N. (1997). The pruning of fruit trees and vines. Kalyani Pub., New Delhi-Ludhiana, pp. 124-126.

Bal, J.S. 2006. Fruit Growing : Fruit plant management practices. Kalyani Publishers., pp. 54-66.

Banik B C and Sen S K. 2002. Effect of pruning and fertilization on growth and yield of mango cv. Fazli. *Orissa Journal of Horticulture*. **30**(1):13-18.

Barritt B H. 1992. Hybrid tree cone orchard system for apple. *Acta Horticulturae* **322**: 87-92.

Basu J, Das B, Sarkar S, Mandal K K, Banik B C, Kundu S, Hassan M A, Jha S and Ray S K.2007. Studies on the response of pruning for rejuvenation of old guava orchard. *Acta Horticulturae* **735**:303-9.

Batten D J and Lavah E. 1994. Base temperature for growth processes of lychee, a recurrently flushing tree are similar but optima differ. *Australian Journal of Plant Physiology* **21**:589-602.

Batten D J, McConchie C A and Llyod J. 1994. Effects of soil water deficit on gas exchange characteristics and water relations of orchard litchi (*Litchi chinensis* Sonn.) trees. *Tree Physiology* **14**:1177-89.

Batten, D J and McConchie C A. 1995. Floral induction in growing buds of lychee (*Litchi chinensis*) and mango (*Mangifera indica*). *Aust. J. of Plant Physio.*, **22**: 783-791.

Bell A D.1991. Plant Form, Oxford University Press, New York.

Burondkar M M, Gunjate R T, Megdum M B and Govekar M A. 2000. Rejuvenation of old and over crowded Alphanso mango orchard with pruning and use of paclobutrazol. *Acta Horticulurae* **509**:681-686.

Cary, P.R. 1981. Citrus tree density and pruning practices for the 21st century. Proc. Int. Soc. *Citriculture* **1**:165-168.

Castle, W.S. 1978. Controlling citrus tree size with rootstocks and viruses for higher density plantings. *Proc. Fla. State Hort. Soc.* **91**:46-50.

Chadha K L, Singh A K, Singh S K and Dhillon W S. 2013. Horticulture for Food and Environment Security, pp 644, Westville Publishing House, New Delhi.

Chadha K L. 2003. Ber. In : *Handbook of Horticulture*. DIPA (ICAR) Publication, New Delhi., pp. 154-157.

Chadha K L. 2003. Guava. *In : Handbook of Horticulture*. DIPA (ICAR) Publication, New Delhi., pp. 189-194.

Chadha K L. 2003. Jackfruit. *In : Handbook of Horticulture.* DIPA (ICAR) Publication, New Delhi., pp. 195-196.

Chadha K L. 2003. Pomegranate. *In : Handbook of Horticulture.* DIPA (ICAR) Publication, New Delhi., pp. 297-304.

Chadha K L. 2003. Sapota. *In : Handbook of Horticulture.* DIPA (ICAR) Publication, New Delhi., pp. 308-312.

Chandra K K, Pandey S K and Singh S K. 2012. Influence of tree rejuvenation, IPNM and VA Mycorrhizal fungi on shoot emergence, yield and fruit quality of *Psidium guajava* under farmers field condition. *International Journal of Biosciences.* **2**(11): 9-17.

Chang M C and Lu C J. 1995. Effect of soil water content and development, yield and quality of mango (*Mangifera indica* L.), *Research Bulletin*, Tainan, District. Agricultural Improvement Station, **32**:45-55.

Chaudhuri S. 1949. Some studies in Rhizopus rot of jackfruit. *J. Indian Bot.Sc.* **18**:42-50.

Coley, P. D., J. P. Bryant, and F S. Chapin III. 1985. Resource availability and plant anti-herbivore defense. *Science* **230**: 895-899.

Crane J H and Balerdi C F. 2005. Guava Growing in the Florida Home Landscape. Horticultural Sciences Department document HS4. Florida Cooperative Extension Service, Institute of Food and Agricultural Sciences, University of Florida. pp. 1-8.

Crane J H, Garcia L Salazor, Lin T S, Pinto A C de Queiroz and Shu Z H. 2009. Crop production management. *In: The Mango (2nd Edition): Botany, Production and Uses* pp 432-83 Richard E. Litz (ed), CABI Publication, UK.

Cutting J G M and Bower J P. 1990. Relationship between auxin transport and calcium allocation in vegetative and reproductive flushes in avocado. *Acta Horticulturae* **275**: 469–476.

Datta M and Dhiman K R. 2001. Effect of some multipurpose trees on soil properties and crop productivity in Tripura. *J. Indian Soc. Soil Science.* **49**(3):511-515.

Datta M. Laskar S and Prasad R N. 1995. Effect of land use system on physiochemical characteristics and nutritional status of soil in Tripura. *Indian J. Hill Farming.* **8**(1):22-27.

Davenport T L. 2006. Pruning strategies to maximize tropical mango production from the time planting to restoration of old orchards. *Hort. Science* **41**(3):544-548.

Davenport T. 1993. Floral manipulation in mangos. *In : Proceedings of a Conference on Mango in Hawaii,* pp 54-60,University of Hawaii at Manoa, College of Tropical Agriculture and Human Resources, Honolulu, Hawaii.

Davenport T.L. 2006. Pruning strategies to maximize tropical mango production from the time planting to restoration of old orchards. *Hort. Science* **41**(3):544-548.

Dodd J C, Prusky D and Jeffries P. 1997. Fruit diseases. *In: The Mango: Botany, Production and Uses*, Litz R E (ed.) CABI Publication, Wellington, UK. pp.281-325,

Dora D K and Sethi Kabita. 2013. Canopy management in cashew : A lucrative venture. *In : Canopy management and high density planting in sub-tropical fruit crops*, pp 140-7, V.K. Singh and H. Ravishankar (Eds.). *Proceedings of National Seminar cum Workshop*. CISH, Lucknow.

Durland G. 1997. Effects of light availability on the architecture of canopy in mango (*Mangifera indicia* L.) cv. Manzana trees. *Acta Horticulture* **455**: 217-272.

Dwivedi S K and Dwivedi P.1999. Wilt disease of guava: a national problem. *J. Applied Hort.* **1**(2):151-154.

Fakhri A. Bazzaz, Nona R. Chiariello, Phyllis D. Coley, Louis F. Pitelka. 1987. How Plants Cope: Plant Physiological Ecology, *BioScience*, **37**(1): 58-67.

Flore J A. 1994. Stone fruits in handbook of environmental physiology of fruit crops, *Vol. I: Temperate Crops,* Skefer B and Anderson P C (Eds.), CRC Press, Bokarotan, UK. pp. 230-70.

Flore J A. Kesner C D and Webster A D. 1996. Tree canopy management and the orchard environment: principles and practice of pruning and training. *In: A D Webster, and N E Looney* (Eds.), *Cherries Crop physiology, production and uses*, Wallingford, Oxon, UK: CAB International, p. 513.

Galan Sauco, V. and Menini, U.G. 1989. Lychee cultivation. FAO Plant Production and Protection Paper No. 83. Food and Agriculture Organization of United Nationas, Rome, p. 205.

Gaur, A.C., Neelkanthan, S., Dargan, K.S. 1971. Organic manures. ICAR Publication, New Delhi, pp. 3-10.

Gill K S and Bal J S.2006.. Influence of pruning severity and time on yield and fruit quality of ber (*Zizyphus mauritiana L.*) cv. Umran. *Indian J.Hort.* **63** (2): 162-165.

Go´mez-Cadenas A, Mehouachi J, Tadeo FR, Primo-Millo E, Talon M. 2000. Hormonal regulation of fruitlet abscission induced by carbohydrate shortage in citrus. *Planta* **210**: 636–643.

Gonzalez J L, Foguet J L, Blanco A S, Vinciguerra H F and Glencross S. 1996. Rejuvenation of organce trees by pruning. *Advance Agroindustrial* **17**(67):3-6.

Goren M and Gazit S. 1992. Small-Statured Litchi Orchards; A New Approach to the Growing of Litchi. *Acta Horticulturae* **349**:69-72.

Gurupasand T R, Hanumanthappa M, Kamath K, Sudhir V and Dhananjaya B. 2012. Technologies for improvement of senile cashew garden. *Challenges and Opportunites*, pp. 35-9.

Hackett W P. 1985. Juvenility, maturation and rejuvenation in woody plants. *Horticultural Reviews,* **7**:109-55.

Halle F. Oldeman RAA and Tomlinson P B. 1978. Tropical trees and forests. An architectural analysis, Springer Verlag, Berlin-Hiedelberg.

Hare Krishna. 2012. Physiology of fruit production. Studium Press (India) Pvt. Ltd., pp 101.

Hassan M A, Singh B, Sarkar S, Jha S and Ray S K. 2009. Canopy management of unproductive mango (*Mangifera indica* L.) orchards. *Acta Horticulturae* **820**: 339-346.

Hickman, J. C. 1977. Energy allocation and niche differentiation in four co-existing annual species of Polygonum in western North America. *J. Ecol.* **65**: 317-326.

Hieke S, Menzel CM, Ludders P. 2002b. Shoot development, chlorophyll, gas exchange and carbohydrates in lychee seedlings (*Litchi chinensis*). Tree Physiology **22**: 947–953.

Hieke S. Menzel C M, Doogan V J and Ludders P. 2002. The relationship between fruit and leaf growth in lychee (*Litchi chinensis* Sonn.) *Journal of Horticultutral Science and Biotechnology* **77** (3):320-325.

Hiwale S S. 2006. Rejuvenation of pomegranate orchards through non-selective pruning. *Proceedings of the National Symposium on Production, utilization and export of underutilized fruits with commercial potentialities*, pp.174-6, Kalyani, WB,

Hiwale S S. 2009. Orchard rejuvenation. *In: The Pomegranate*, pp. 67-72, New India Publishing Agency, New Delhi.

Hiwale, S.S. and Raturi, G.B. (1993). Effect of pruning severity on growth, yield and quality of ber. *prog. Hort.* **25** (3/4): 161-163.

Huang Y H, Su J F and Wang Z X. 1992. Discussion on the relationship between longan tree structure and yield. *Fujian Agricultural Science and Technology*. **1**:17-18.

Hubballi V N and Jnanadevan R. 2012. Cashew development in India : Strategies for sustainable production. *In : International conference of sustainable cashew production – challenges and opportunities*, pp. 17-27, Goa.

Huett D O. 1996. Prospects for manipulating the vegetative–reproductive balance in horticultural crops through nitrogen nutrition: a review. Australian Journal of Agricultural Research **47**: 47–66.

Hung Nguyen QUOC 2011. Jackfruit in Vietnam. *In: The Jackfruit* (Ed Sudha George Valavi, K V Peter and George Thottappilly). Studium Press LLC.USA. pp. 463-474.

ICAR-DataBook-2014. Published by Director, Indian Agricultural Statistics Research Institute, New Delhi., p. 146.

Intrigliolo F. 1984. Citrus pruning. *Informatore Agrario*, **40**(22): 49-53.

Iyer C P A and Kurian Reju M. 2006. High density planting in tropical fruits-Principles and practices. International Book Distributing Co., Lucknow.

Jackson J E. 1980. Light interception and utilization by orchard system, *Horticultural Review* **2**:208-287.

Jahangeer A, Baba P, Akbar Ishfaq and Kumar V. 2011. Rejuvenation of old and senile orchards: A review. Hi-Tech Horticultural Society. *Annals of Horticulture.* **4**(1): 37-44.

Janik J and Paull R E. 2008. The Encylopedia of Fruits and Nuts. *In Anacardiaceae – Cashew nut* (*Anacardium occidentale* L.), CABI Publication, Wellington, UK. pp 8-14,

Jauhari Q S and Singh R D. 1971. Bael –valuable fruit. *Indian Hort.* **16**:9-10.

Kahlon G S. 2008. Citrus industry – Rehabilitating in north India. *In : Recent Initiatives in Horticulture* (Eds K L Chadha, A K Singh, and V. B. Patel)., p 253-261.

Kalloo G, Reddy B M C, Singh G and Lal B. 2005. Rejuvenation of old and senile orchards. *Bulletin,* pp 40, CISH, Lucknow.

Karasek R A and Theorell T. 1990. Healthy work: stress productivity and the reconstruction of working life. Basics books, New York.

Khan M and Syamal M M. 2004. Effect of pruning intensity on flowering and fruiting of Kagzi lime (*Citrus aurantifolia* Swingle) *Indian J. Hort.* **61** (2) : 171-172.

Koo R C J. and Muraro R P. 1982. Effect of tree spacing on fruit production and net returns of 'Pineapple' oranges. *Proc. Fla. State Hort. Soc.* **95**:29-33.

Kriedemann P E. 1986. Tree water relations. *Acta Horticultureae.* **175**: 343-350.

Kumar D P, Hegde M and Guruprasad T R. 1998. Fertilisation for higher nut production in cashew. *Cashew Bull.* **35**:2.

Kumar Rajesh. 2007. Bigde bagon ki raunak wapas aye, *Phal Phul* **6**(4):23-25.

Kumar Rajesh. 2012. Adoption of improved production technologies by litchi growers for quality production. *Journal of Community Mobilization and Sustainable Development* **7**(2):210-6.

Kumar Kailash and Adak Tarun. 2013. Integrated nutrient and water management. *In : Recent advances in subtropical fruit crop production,* pp 40-50, Singh V K and Ravishankar H (Eds.), CISH, Lucknow Publication.

Kumar N and Kavino M. 2012. Canopy management in Mango. *In : Mango (Volume – 1) : Production and processing technology*, pp366-383, Valavi Sudha G, Rajamohan K, Govil J N, Peter K V and Thottappilly George (Eds.), CABI Publications, UK.

Kumar R. 2013. Effect of vegetative flushing and shoot maturity on flowering, bearing behaviour and fruit yield in litchi (*Litchi chinensis*). *Current Horticulture* **1**(1): 30-34.

Kumar Rajesh. 2008. Rejuvenaqion of old senile litchi orchards, National Research Centre on Litchi, *Technical Bulletin* **5**: 1-8.

Kumar Rajesh. 2009. Transformation of old senile litchi orchards for quality fruit production. *In: Souvenir*, pp 51-4, Swadesh Prem Jagriti and Sangosthi-2009, Trivedi G, Singh R K and Sharma D (Eds), Lt Amit Singh Memorial Foundation, New Delhi.

Kumar Rajesh. 2013. Reiterative pruning and canopy rebuilding in old senile litchi trees for rejuvenation enhanced quality fruit production. *In : Canopy management and high density planting in sub-tropical fruit crops,* pp 217-224, Singh V K and Ravishankar H (Eds.). *Proceedings of National Seminar cum Workshop,* CISH, Lucknow.

Kumar Rajesh. 2013. Rejuvenating old senile litchi trees- A remunerative venture. *In : Canopy management and high density planting in sub-tropical fruit crops,* Singh V K and Ravishankar H (Eds.). *Proceedings of National Seminar cum Workshop,* pp 264-9, CISH, Lucknow.

Kumar, M. and Lal, R.L. 2007. Canopy management by pruning under varying spacing-cum-planting systems in litchi (*Litchi chinensis* Sonn.). *Pantnagar J. Res.* **5**(1): 45-49.

Kundu S S, Pareek O P and Gupta A K. 1995. Effect of time and severity of pruning on physico-chemical characteristics of ber (*Zizyphus mauritiana* Lamk.). *Haryana J. Hort.Sci.* **24** (1): 53-58.

Lal B and Mishra D. 2007. Effect of pruning on growth and bearing behavior of mango cv. Chausa. *Indian Journal of Horticulture* **64**(3): 268-270.

Lal B and Mishra D. 2008. Studies on pruning in mango for rejuvenation. *Indian Journal of Horticulture* **64**(4): 405-408.

Lal B, Rajput M S and Rajan S. 2001. Rejuvenation of old and unproductive mango orchards, *Indian Horticulture*, **46**(2):24-25.

Lal B, Rajput M S, Rajan S and Rathore D S. 2000. Effect of pruning on rejuvenation of old mango trees. *Indian Journal Horticulture*, **57**(3):240-242.

Larson K D, Schaffer B and Davies F S. 1989. Effect of irrigation on leaf water potential, growth and yield of mango trees. *Proceedings of the Florida State Horticulture Society* **102**:226-8.

Lefebvre, A. 1973. Mineral nutrition of the cashew. *Fruits* **28** (10): 691-697.

Lenka P C and Pattnaik A K. 2008. Cashew – Research and Development. *In: Recent Initiatives in Horticulture* (Eds K L Chadha, A K Singh and V. B. Patel)., pp. 266-274.

Lewis L N. C D McCarty and P W Moore.1963. Effects of skeletonization on the rejuvenation and production of declining orange trees. *Proceedings of the American Society for Horticultural Science* **83**: 295–299.

Maity R G and Das P C. 2004. *Fruit crops of India*. Kalyani Publishers, New Delhi.

Maity, S.C. and Mitra, S.K. 2001. Litchi. In: T.K.Bose, S.K.Mitra and D.Sanyal (eds.), Fruits: Tropical and subtropical. Vol. 1 Naya Udyog,Calcutta, India, pp. 556-608.

Mallik C P and Srivastava A K. 1985. Textbook of Plant Physiology. Kalyani Publishers, New Delhi.

Masarrat H, Abbas S R, Srivastava R P and Sharma S. 2000. Studies on insect pests of aonla (*Embelica officinalis* Linn.). *Annals of Plant Protection Sciences*. **8**:85-8.

McConchie C A and Batten D J. 1991. Fruit set in lychee (*Litchi chinensis*). Variation between flowers, panicles and trees. *Australian Journal Agricultural. Research* **42**:1163-72.

Mehta S S and Srivastava Manish 2008. Aonla-Commercialization in India. *In : Recent Initiatives in Horticulture* (Eds K L Chadha, A K Singh, and V. B. Patel)., pp.288-296.

Menzel C M and Simpson D R. 1990. Nutritional studies on lychee trees in subtropical Australia. *Acta Horticulturae* **273**:581-5.

Menzel C M, Olesan T, McChonchie C, Wiltshire N, Diczabilis Y and Wicks C. 2000. Lychee, longan and rambutan-Optimizing canopy management. *A report for the Rural Industries Research and Development Corpotation* (RIRDC), RIRDC Publication No. 00/29, RIRDC Canberra.

Menzel, C.M. 1987. Lychee nutrition; a review. *Scientia Horticulturae* **3**(3/4):195-224.

Menzel, C.M. and Waite, G. K.2005. Litchi and longan: Botany, production and uses. CABI Publishing. pp. 88-166.

Menzel, C.M., Rasmussen,T.S. and Simpson,D.R. 1995. Carbohydrate reserves in lychee trees (*Litchi chinensis* Sonn.*). Journal of Horticultural Science* **70** (2):245-255.

Mika A. 1986. Physiological responses of fruit trees to pruning. *Horticultural Reviews*. **8**:337-78.

Milan Abd Rahman. 2011. Jackfruit in Malaysia. *In: The Jackfruit* (Ed Sudha George Valavi, K V Peter and George Thottappilly). Studium Press LLC.USA.pp. 445-462.

Mishra Alok and Kaushal Krishna Kumar. 2013. Government initiatives on canopy management in horticulture. *In : Canopy management and high density planting in sub-tropical fruit crops*, pp 19-22, Singh, V K and Ravishankar H (Eds.). *Proceedings of National Seminar cum Workshop*. CISH, Lucknow.

Mishra D, Pandey D, Mishra R and Pathak R K. 2007. Performance of improved aonla cultivars during top-working of senile trees. *Indian Journal of Horticulture* **64**(4): 396-398.

Mishra D. 2013. Improving orchard productivity through rejuvenation and canopy management. *In : Recent advances in subtropical fruit crop production* Singh, V K and Ravishankar H (Eds.), CISH, Lucknow.

Mishra, S., Choudhary, M.R., Yadav, B.L. and Singh, S.P. 2011. Studies on the response of integrated nutrient management on growth and yield of ber. *Indian J. Hort.* **68**: 318-321.

Misra K K and Uniyal Shweta.2013. Bael. *In: Fruit Production in India.* (Ed. W S Dhillon), Narendra Publishing House, Delhi, pp. 83-94.

Mistry P M and Patel B N. 2009. Impact of heading back plus paclobutrazol on rejuvenation of old and overcrowded Alphonso orchards. *Indian Journal of Horticulture,* **66**(4):520-1.

Mitra S K and Dutta R S K.2005. Advances in the production and research of lychee in India. *Acta Horticultureae,* **665**. 47-51.

Mitra S K and Sanyal D. 2004. Guava. ICAR (DIPA) Publications., pp. 1-64.

Mitra S K, Gurung M R. and Pathak P K. 2008. Guava production and improvement in India: an overview. *Acta Horticulture* **787**: 59-65.

Mitra S K.2002. Overview of lychee production in the Asia-Pacific region. Lychee production in the Asia -Pacific region.FAO Publication : 5-13.

Morton I P. 1960. The emblica (*Phyllantus emblica* L.). A rich but neglected source of Vitamin C. *Proc. Flor. State Hort. Soc.* **68**:315-321.

Muthulakshmy P. 2003. Genetic diversity and canopy management in jackfruit (*Artocarpus heterophyllus* Lam.) PhD. (Hort.) thesis submitted to Kerala Agril.Univ. p.231.

Nalini P V, Nair G S and Babu L C. 1999. Canopy characteristics in cashew (*Anacardium occidentale* L.) *The Cashew*, **13**:31-36.

Nath V, Kumar Dinesh and Pandey V. 2008. Fruits for the future (Vol. I) : Well versed arid and semi arid fruits. SSPH, Delhi.

Nath, Vishal, Das Bikash, Rai, Mathura and Rai, Rajesh Kumar. 2003. Center opening induces stereo fruiting in litchi cv. Shahi. *Progressive Horticulture*, **35**(2):219-220.

Nayak M G. 2011. Canopy management and rejuvenation of senile orchard in cashew. *Presentation made in National Expert Consultation on cashew : Challenges and strategies for higher productivity and quality of cashew,* (23-24 December), Puttur, Karnataka.

NHB 2007. Mango status in Indian Horticulture Database, National Horticulture Board, Gurgaon, Haryana.

Nimbalkar C A, Idate G M, Gaikwad S P and Desai U T. 2010. Rejuvenation technology for improvement of guava (cv Sardar) orchard. *Journal of Maharashtra Agricultural Universities*, **35**(3):486-7.

NRCL. 2014. Annual report, National Research Centre on Litchi, Muzaffarpur, Bihar.

Olesan T, Menzel C M, Wiltshire N and McConchie C A. 2002. Flowering and shoot elongation of lychee in Eastern Australia. *Australian Journal of Agricultural Research.* **53**:977-983.

Oostrum A Van. 1962. Rejuvenate citrus trees by pruning. *Farming in South Africa.* No. **46**. Osmond and H. Ziegler) Springer-Verlag, Berlin, pp. 233-280.

Pal Mahendra and Misra K K 2005. Effect of pollen on fruit set, fruit drop and final fruit retention in self and cross-pollination in bael (*Aegle marmelos*). *Indian J Agri. Sci.* **75**(12):800-804.

Pandey D, Pandey A and Yadav Sarita K. 2013. Horticultural interventions for increasing productivity of aonla and bael. *In : Recent advances in subtropical fruit crop production,* pp. 76-86. Singh, V K and Ravishankar H (Eds.), CISH, Lucknow.

Pathak R K. 2003. Status report of genetic resources of Indian Gooseberry – Aonla (*Embelica officinalis* L.) in South East Asia. IPGRI Office, New Delhi.

Patil Prakash.2013. Jackfruit. *In: Fruit Production in India.* (Ed. W S Dhillon), Narendra Publishing House, Delhi, pp. 19-30.

Phillips, R.L. 1974. Performance of 'Pineapple' orange at three tree spacings. *Proc. Fla. State Hort. Soc.* **87**:81-84.

Phillips, R.L. 1978. Hedging and topping citrus in high-density plantings. *Proc. Fla. State Hort. Soc.* **91**:43-46.

Ploetz R C and Prakash O. 1997. Foliar, floral and soil borne diseases. *In: The Mango: Botany, Production and Uses,* pp. 281-325, Litz R E (ed.) CABI Publication, Wellington, UK.

Ploetz R C. 2003. Diseases of Mango. *In: Diseases of Tropical Fruit Crops,* pp. 327-63, Ploetz R C (ed.), CABI Publication, Wellington, UK.

Rai M, Nath Vishal and Dey P. 2001. Litchi, CHES, Plandu, Ranchi.

Ram S, Dwivedi T S and Bist L D. 1977. Internal fruit necrosis in aonla (*Emblica officinalis* Gaertn). *Prog.Hort.* **8**:5-12.

Rao V N M and Shanmugavellu K G. 1976. Studies on the effect of pruning in mango. *Progressive Horticulture* **8**:21-28.

Rao V N M and Shanmugavellu K G. 1975. Mango respond to pruning. *Indian Horticulture* **20**(3): 3-5.

Rao V N M. 1971. A note on pruning as a remedy for irregular bearing mango. *Andhra Agricultural Journal* **18**: 242-245.

Reddy B M C and Patil Prakash. 2004. Training and Pruning. *In: Advances in Citriculture,* pp 206-19, Singh Shyam, Srivastava V J, Srivastava A K and Singh I P (Eds.). Jagmender Book Agency Publication, New Delhi.

Reddy R T N and Kurian R M. 2011. Studies on rejuvenation of old, unproductive 'Alphonso' mango trees in orchards.

Room P M. Maillette L and Hanan J S. 1994. Modular and metamer dynamics and virtual plants. *Advances in Ecological Research*, **25**:105-57.

Salazar D M, Melgarejo P, Martinez R, Martinez J J, Hernandez F and Burguera M. 2006. Phenological growth stages of guava tree (*Psidium guava* L.). *Scientia Horticulturae* **108**: 157-161.

Sandhu Savreet and Gill B S. 2013. Pomegranate. In: *Physiological disorders in fruit crops*. New India Publishing Agency, New Delhi, pp. 171-176.

Singh A K and Bajpai Anju.2013. Guava. *In: Fruit Production in India*. (Ed. W S Dhillon), Narendra Publishing House, Delhi, pp. 263-278.

Singh A K, Uniyan Shweta and Misra K K. 2013. Aonla. In: Fruit Production in India. (Ed. W S Dhillon), Narendra Publishing House, Delhi, pp. 19-30.

Singh Amar. 2003. Fruit physiology and production, pp 488, Kalyani Publishers, New Delhi.

Singh B. 2007. Rejuvenation of old orchard. *In : Horticulture at a glance*. Kalyani Publishers. pp. 43-5.

Singh G, Kumar R and Vishal Nath. 2011. Rejuvenation in litchi. NRC on Litchi, Muzaffarpur, Bihar, pp. 1-14.

Singh G, Mishra R and Gupta S. 2007. Modifying existing Guava tree canopies for increased production efficiency. *Acta Horticulturae,* **735**: 243-248.

Singh G, Mishra R and Singh G P. 2005. Guava rejuvenation. Central Institute of Subtropical Horticulture, Lucknow. pp. 1-20.

Singh G. 2005. Strategies for improved production of guava. Souvenir, First International Guava Symposium, Lukhnow, India, pp. 26-39.

Singh G. 2008. Guava-Meadow orcharding. *In: Recent Initiatives in Horticulture* (Eds K L Chadha, A K Singh, and V. B. Patel)., pp. 261-265.

Singh G. 2010.Canopy management-An essential and integral element for healthy orchard establishment and sustainable production. *Prog. Horti.* **42** (1): 15-17.

Singh Gorakh and Mishra Rajneesh. 2007. Aonla rejuvenation. PDFC, CISH, Lucknow, pp. 1-16.

Singh Gurdarsha and Chanana Y R. 2005. Influence of pruning intensity on and pruning frequency on vegetative and reproductive attributes in guava cv. L-49. *First International Guava Symposium,* Central Institute of Subtropical Horticulture, Rehmankhera, Lucknow, p. 52.

Singh I S. 2012.Aonla: Production, Handling and Processing. Westville Publishing House, New Delhi. pp. 1-166.

Singh L B.1953. Top working of aonla (*Phyllanthus emblica*) *Indian J. Hort.* **10**:59-61.

Singh N K, Srivastava D C and Bhandarkar A P. 2012. Growth, yield and quality of guava as influenced by varying rejuvenation periods. *Indian Journal of Horticulture*, **69**(2):181.

Singh S K. 2005. Studies on pruning behavior in Aonla (*Embilca officinalis* Garten.) cv. NA-7 NDUAT thesis Kumarganj, Faizabad, (U.P).

Singh S S, Vishal Nath, Singh Abha and Pandey S D. 2012. Litchi : Preventive practices and curative measures, SSPH, New Delhi.

Singh Shyam, Srivastava V J, Srivastava A K and Singh I P. 2004. Advances in Citriculture. Jagmender Book Agency Publication, New Delhi.

Singh Sukhdev, Singh Bikramjit and Singh Manwinder.2010. Influence of time and severity of pruning on vegetative growth, fruit quality and yield of ber cv. Umran. National Seminar on recent trends and development, CCS. HAU. Hissar, pp. 57-58.

Singh V K and Ravishankar H. 2013. Canopy management and high density planting in sub-tropical fruit crops. *Proceedings of National Seminar cum Workshop,* CISH, Lucknow.UP.

Singh V K and Singh G. 2007. Photosynthetic efficiency, canopy micro-climate and yield of rejuvenated guava trees. *Acta Horticulturae*, **735**: 249-257.

Singh V K, Soni M K and Singh A. 2013. Strategies and prospects of high density planting and canopy management in guava (*Psidium guajava* L.) *In : Canopy management and high density planting in sub-tropical fruit crops - Proceedings of National Seminar cum Workshop,* pp 23-33, Singh V K and Ravishankar H (Eds.),CISH, Lucknow.

Singh V K. 2013. Factors affecting productivity with reference to alternate bearing and other physiological disorders in mango (*Mangifera indica* L.). *In : Recent advances in subtropical fruit crop production,* Singh, V.K and Ravishankar, H. (Eds.), CISH Lucknow pp. 18-25.

Singh, Amar. 2006. Horticulture at a glance: Fruits. Kalyani Publishers, pp. 76-99.

Singh, B. 2007. Rejuvenation of old orchard. *In: Horticulture at a glance*. Kalyani Publishers. pp. 43-45.

Singh, G. 2010. Pomegranate. *In: Practical manual on canopy management in fruit crops*. Published by Department of Agriculture and Cooperation, Ministry of Agriculture, Govt. of India, New Delhi, pp. 71-72.

Singh, G. 2010. Canopy management-An essential and integral element for healthy orchard establishment and sustainable production. *Prog. Horti.* **42** (1):15-17.

Singh, Gorakh. 2010. Canopy management in fruit crops, pp. 51-53.

Singh, H., and Bal, J.S. and Singh, G (2004). Standarization of pruning techniquein ber. A Review. *Indian J. Hort. Sci.*. **61** (3): 259-260.

Singh, H.P. and Babita, S.2002. Lychee production in India. Lychee production in the Asia -Pacific region.FAO Publication, pp. 55-77.

Singh, I.P. and Singh, S. 2003. Exploration, collection and mapping of citrus genetic diversity in India. *Technical Bulletin No.* 7, National Research Centre for Citrus, Nagpur, Maharashtra, India.

Singh, Rajbir and Bal, J.S. (2008). Pruning in ber (*Zizyphus mauritana* Lamk.) *Agric. Rew.*, **29** (1): 61-67.

Singh, S., Shivankar, V.J., Gupta, S.G., Singh, I.P., Srivastava, A.K. and Das, A.K. 2006. Citrus in NEH region. National Research Centre for Citrus Publ., Nagpur, Maharashtra, India, pp. 1-179.

Skefer B and Gey G O. 1989. Seasonal effects of pruning on light penetration, specific leaf density and chlophyl content of mango, *Scientia Horticulturae* **41**:55-61.

Soepadmo E. 1992. *Artocarpus heterophyllus* Lam. In Verheij E W M and Coronel R E(eds), Plant Resources of South-east Asia, No.2. Edible Fruit and Nuts. PROSEA, Wegeningen, Netherlands, pp. 86-91.

Sonwane, P.C. and Desai, U.T. 1989. Performance of staggered cropping in Pomegrante. *J. Maharashtra Agric. Univ.* **14**(3):341-342.

Srivastava K K. 2007. Canopy management in fruit crops. International Book Distributing Co., Lucknow.

Srivastava A K and Singh S. 2004. Citrus nutrition. *In: Advances in Citriculture,* pp 305-62. Singh, Shyam, Srivastava V J, Srivastava A K and Singh I P. (Eds.), Jagmender Book Agency, New Delhi.

Srivastava A K and Singh S. 2008. Citrus decline : soil fertility and plant nutrition. *Journal of Plant Nutrition.* **32**(2):197-245.

Srivastava A K. 2011. Development of INM to harmonize with improved citrus production: Changing scenario. *In: Plant Physiology under Changing Environment,* pp 294-368, Hemantarajan A (Ed), Scientific Publishers (India), Jodhpur.

Srivastava A K. and Singh S. 2007. Nutrient management in citrus orchards of northeast India. *Indian Farming* **59**(6): 27-32.

Srivastava, A.K. and Singh, S. 2005. Diagnosis of nutrient constraints in citrus orchards of humid tropical India. *Journal of Plant Nutrition* **29**(6): 1061-1076.

Stassen, P J C, Grove H G and Davie S J. 1999. Tree shaping strategies for higher density mango orchards. *Journal of Applied Horticulture* **1**(1):1-4.

Stassen, P.l.C. and Davie, S.J. (1996). Principles of a Tree Management Programme. Institute for Tropical and Subtropical Crops Information Bulletin No. 285.

Stebbins R L. 1985. Pruning : How to Guide for gardeners. HP Books, Price Stern Sloan, Losangeles, California.

Suharto I, Ambarawati I G A A, Agung I G A M S and Nurjaya I G M O. 2012. The number of grafted scions and remaining productive branches after new shoot growth and flowering of side grafted cashew (*Anacardium occidentale* L.). *Journal of International Society for South east Asian Agricultural Science*. **18**(1):160-172.

Tangonan Naomi G. 2011. Diseases of Jackfruit and their management. *In: The Jackfruit* (Ed Sudha George Valavi, K V Peter and George Thottappilly). Studium Press LLC.USA.pp. 197-205.

Tucker D P H and Wheaton T A. 1978. Trends in higher citrus planting densities. *Proc. Fla. State Hort. Soc.* **91**:36-40.

Tyagi R M S and Pathak R K.1987. Control of Aonla (*Emblica officinalis* Gaerth.) rust (*Ravenelia emblicae* Syd.) through fungicide. *Indian J Mycol. Pl. Pathol.* **17**(2): 218-219.

Valavi Sudha George, Jacob B, Amin Muhammad Nurul, Magdalita Pablito M and Thottappilly George. 2011. Harvesting, Grading, Marketing and Economics of Jackfruit production. *In: The Jackfruit* (Ed Sudha George Valavi, K V Peter and George Thottappilly). Studium Press LLC.USA. pp. 351-365.

Veeraraghavan P G, Celine V A and Balakrishnan S. 1985. Study on the fertilizer requirements of cashew (*Anacardium occidentale* L.). *Cashew Causerie* **7**(2):6-8.

Wali V K, Bakshi P, Sharma A, Singh A P, Bakshi M and Kour N. 2013. Rejuvenation of Unproductive Old Mango Orchards, Booklet, Division of Fruit Science, *SKUAST-J.* p. 32.

Wasker D P. 2008. Pomegranate: Present status, constraints and opportunities. *In : Recent Initiatives in Horticulture*. K. L. Chadha, A.K.Singh and V.B.Patel (Eds). Horticulture Society of India, New Delhi, pp. 417-432.

Wheaton T A, Castle W S, Tucker D P H and Whitney J D. 1978. Higher density plantings for Florida citrus - concepts. *Proc. Fla. State Hort. Soc.* **91**:27-33.

Wheaton T A, Castle W S, Whitney J D, Tucker D P H and Muraro R P. 1990. A high density citrus planting. *Proc. Fla. State Hort. Soc.* **103**:55- 59.

Wheaton T A, Whitney J D, and Tucker D P H. 1986. Tree spacing and rootstock affect growth, yield, fruit quality, and freeze damage of young 'Hamlin' and 'Valencia' orange trees. *Proc. Fla. State Hort. Soc.* **99**:29-32.

Wheaton T A, Whitney J D, Tucker D P H and Castle W S. 1984. Cross hedging, tree removal, and topping affect fruit yield and quality of citrus hedgerows. *Proc. Int. Soc. Citriculture* **1**:109-114.

Whiley A W, Rasmussen T S, Saranah J B and Wolstenholme B N. 1989. Effect of temperature on growth, dry matter production and starch accumulation in ten mango (*Mangifera indica* L.) cultivars. *Journal of Horticultural Science* **64**: 753–765.

White J. 1979. The plant as a meta-population. *Annual Review of Ecology and Systematics*, **10:**109-45.

Whitney J D and Wheaton T A. 1984. Tree spacing affects citrus fruit distribution and yield. *Proc. Fla. State Hort. Soc.* **97**:44-47.

www.IndianJournals.com : Rejuvenation of old and Senile Orchards: A Review, Downloaded From IP - 144.16.122.1 on dated 6-Jun-2012.

Yadav A K, Singh S P and Singh Alok. 2005. Screening of ber (*Zizyphus mauritiana* Lamk.) cultivars based on external, physical and chemical attributes for cultivation in Eastern Uttar Pradesh. *Horticultural Journal.* **18**(3).144-148.

Yuan R C and Huang H B. 1988. Litchi fruit abscission : its patterns effect of shading and relation to endogenous abscisic acid. *Scientia Horticulturae,* **36**:281-292.

Zhang, Z .W. 1997. China, the native home of litchi. *In: Litchi: Pictorial Narration of Cultivation* (Eds..Zhang, Z.W., Yuan, P.Y.,Wang, B.Q., Qui, Y.P. and Li, J.S.). Pomology Research Institute, Guangdong Academy of Agricultural Science, China: 12-17.

Zheng, Q., Davenport, T.L. and Li, Y.C.2001. Stem age, winter temperature and flowering of lychee in South Florida. *Acta Hortic.***558**: 237-240.

FSC
www.fsc.org
MIX
Papier | Fördert gute Waldnutzung
FSC® C083411

Zeitfracht Medien GmbH
Ferdinand-Jühlke-Straße 7
99095 Erfurt, Deutschland
produktsicherheit@kolibri360.de